Ensuring National Government Stability After US Counterinsurgency Operations

Ensuring National Government Stability After US Counterinsurgency Operations

The Critical Measure of Success

Dallas E. Shaw Jr.

Rapid Communications in Conflict and Security Series
General Editor: Geoffrey R.H. Burn

Amherst, New York

Copyright 2019 Cambria Press

All rights reserved.

No part of this publication may be reproduced, stored in or introduced into a retrieval system, or transmitted, in any form, or by any means (electronic, mechanical, photocopying, recording, or otherwise), without the prior permission of the publisher.

Requests for permission should be directed to:
permissions@cambriapress.com, or mailed to:
Cambria Press
University Corporate Centre,
100 Corporate Parkway, Suite 128
Amherst, New York 14226, USA.

Front cover image: Baghdad, Iraq (Aug. 20, 2006) - US Army soldiers assigned to the 506th Regimental Combat Team, 101st Airborne Division assist the 6th Iraqi Army Division and Iraqi Police to secure the streets for Iraqi citizens. Iraqi citizens were observing the death of the 7th Imam, where one million Iraqi Shias are expected to gather at his burial site in Baghdad to mourn his passing. US Navy photo by Mass Communication Specialists 1st Class Keith W. DeVinney (RELEASED).

Library of Congress Cataloging-in-Publication Data

Names: Shaw, Dallas E., author.

Title: Ensuring national government stability after US counterinsurgency operations : the critical measure of success / Dallas E. Shaw Jr.

Description: Amherst, New York : Cambria Press, 2019. |
Series: Cambria rapid communications in conflict and security series |
Includes bibliographical references and index.

Identifiers: LCCN 2018055674 |

ISBN 9781604979619 (alk. paper)

Hardcover ISBN 9781604979572

Subjects: LCSH: Counterinsurgency--United States--Case studies. | Military assistance, American—Case studies. | Intervention (International law)--Government policy--United States--Case studies. | Combined operations (Military science)--Case studies. | Failed states--Prevention. | Philippines--History--Philippine American War, 1899-1902. | Iraq War, 2003-2011. | Nicaragua. Guardia Nacional--History--20th century. | Vietnam War, 1961-1975. | Nation-building--Case studies.

Classification: LCC U241 .S47 2019 |
DDC 355.02/180973—dc23
LC record available at https://lccn.loc.gov/2018055674

*To my precious bride Jeannette,
The Wife of Noble Character*

Table of Contents

List of Tables ix

List of Figures xi

Acknowledgements xiii

Selective Glossary xv

Chapter 1: Introduction 1

Chapter 2: The Stumbling State 39

Chapter 3: The Strategic Rentier State 75

Chapter 4: The Tumbling State 133

Chapter 5: The Crumbling State 167

Chapter 6: State-centric Counterinsurgency 227

Index 257

Cambria Rapid Communications in Conflict and Security (RCCS)
    Series 275

LIST OF TABLES

Table 1: US COIN Intervention Outcomes Before and After 1950 6

Table 2: Average Monthly US Troop Levels in Iraq War 91

Table 3: Total US Economic and Military Assistance, 2012–2015 93

Table 4: Iraqi Officers and Tactical Disciplines 110

Table 5: Iraqi Officers' Ability to Internalize US Warfighting Concepts 111

Table 6: Iraqi Officer Ability to Internalize US Leadership Traits and Principles 114

Table 7: Growth of the Guardia Nacional, 1927–1932 156

Table 8: Troop Ratios to Population, Terrain, and Adversary 178

Table 9: Growth of the RVNAF, 1954–1967 196

Table 10: Growth of the RVNAF, 1954–1972 199

Table 11: Analysis of Competing Explanations for Increases in SLAW 229

List of Figures

Figure 1: State Longevity Outcomes: Inhabiting Versus Influencing 20

Figure 2: Governance and Security Outcomes Related to Degree of Embeddedness 21

Figure 3: Violent Significant Actions (SIGACTS) 2001–2009 85

Figure 4: Advisors' Perception of Iraqi Competence Upon Withdrawal 108

Figure 5: US Advisor Perceptions of Iraqi Officer Corruption 116

Figure 6: US Advisor Perceptions of Predation Among Iraqi Officers 116

Figure 7: Comparative Guardia Growth, 1927–1933 151

Figure 8: Advisors' Perception of Vietnamese Competence Upon Withdrawal 204

Figure 9: Vietnamese Internalization of US Leadership Traits & Principles 207

Figure 10: US Advisor Perceptions of South Vietnamese Officer Internalization of US Tactical Concepts 209

Figure 11: US Advisor Perceptions of South Vietnamese
Application of Tactical Disciplines 210

Figure 12: US Advisor Perceptions of South Vietnamese Officer
Corruption 211

Acknowledgements

"I will not boast in anything. No gifts, no power, no wisdom. But I will boast in Jesus Christ, His death and resurrection. Why should I gain from His reward? I cannot give an answer. But this I know with all my heart--His wounds have paid my ransom."
—Stuart Townsend,
How Deep the Father's Love for Us

A special thanks goes to my wife Jeannette and Dr. Timothy Ray, a 2015 Distinguished Graduate of the U.S. Army War College, who both provided detailed feedback and edits on this work. I also want to express my gratitude to the Marines, Sailors, and commanders I served with in Iraq and Afghanistan. I especially want to say thank you to my fellow Marines and Sailors in 1st Battalion, 4th Marines and in 2nd Battalion, 9th Marines who died in the service of our nation and their fellow Marines in Sadr City, Hilah, and Ramadi, Iraq and Marjah, Afghanistan. Part of the purpose of writing this book was to help me come to peace with your passing. I miss you all. Semper Fidelis and God Bless.

Selective Glossary

Hamlet Evaluation System (HES): Due to the difficulty associated with determining the success of pacification efforts in South Vietnam, the HES was developed to try to give a quantitative measure of success to pacification efforts. The HES relied on surveys and regularly measured data points such as roads built, school attendance, numbers of violent acts, and other factors to try to provide a numerical value that could be used to assess whether a hamlet was becoming pacified. Critics of the system argued that it was impossible to measure quantitatively that which is inherently subjective. However, proponents of the HES argued the system, as imperfect as it was, allowed decision makers to make more informed decisions about where to invest scarce resources.

Illustrados: These were the wealthy, landowning ethnic Filipino intelligentsia and heads of the ruling feudal system of the Philippines who held power during the US-Philippine War. These wealthy elites formed the Filipino leadership on both sides of the US-Philippine war for both the insurgents and the US-Philippine government.

Lioness Program and Female Engagement Teams: Because of Islamic cultural sensitivities in Iraq and Afghanistan and the prohibition of US men searching Iraqi and Afghan women, the US Marines began the

Lioness program early in the Iraq War. The program requested female Marines to volunteer to serve with male units who engaged with Iraqis on a daily basis. The Lionesses would conduct searches of Iraqi women and teach Iraqi female police officers to conduct searches. The Lioness Program eventually led to the Female Engagement Team program in Afghanistan where female Marines also became responsible for building relationships with Afghan females and gathering information from them in addition to their mission to conduct searches.

Macabebe Scouts: Macabebes were an ethnic minority in the Philippines who had been perennially persecuted by the Tagalog majority. As a result, Macabebes first sided with the Spanish against the Filipino revolutionaries from 1896 to 1898 and then sided with the Americans during the US-Philippine War (1899–1902). They were recognized by the Americans for their bravery and were often praised for being as good as, and even often better than, their American counterparts. They were instrumental in the raid that eventually captured revolutionary leader Emilio Aguinaldo and led to the end of the US-Philippine War. The Macabebe Scouts later became the Philippine Scouts.

Operation Mosthtarak: This was the battle in the Helmand Province of Afghanistan from February to December in 2010. The operation was spearheaded by US Marine infantry working with British and Afghan forces. The initial seizure of Marjah lasted only three weeks, and the operation officially ended in December. However, despite end of the official operation, efforts to secure Marjah and reduce violent significant activity continued until February 2011 when violent significant activity dropped off almost entirely.

Powell-Weinberger Doctrine: It is an amalgamation of criteria announced in 1984 by former Secretary of Defense Casper Weinberger and reiterated by General Colin Powell, then Chairman of the Joint Chiefs of Staff, just before the 1990–1991 US war with Iraq in Kuwait, to enumerate conditions policymakers should agree to before employing US troops overseas. It says, among other things, that the US should only commit troops abroad as a last resort and only when the US intends to win, is willing to employ overwhelming force to do so, has a vital national interest at stake, has clearly articulated goals, and has an exit strategy formulated.

The Surge: The surge was led by General David Petraeus from 2006 to 2008. The surge included an increase to US troop strength in Iraq and a switch from focusing on the destruction of the Iraqi insurgency to the protection of the Iraqi populace. This required US and Iraqi troops to move from their big bases into smaller more distributed combat outposts to have a closer relationship with the local Iraqis. A key aspect in the success of the surge was the cooption of the Sunni tribal leaders in the Al Anbar province and the rise of Awakening Councils. These councils were comprised of influential, grassroots, local leaders and everyday Iraqis. The Sunni Awakening was critical to separating the Iraqi insurgents and particularly Al Qaeda in Iraq from their base of support—the Iraqi people.

Vietnamization: In July 1969, the Nixon Administration announced its policy of gradually Vietnamizing its war in South Vietnam. That is, from the summer of 1969 on, the US would gradually transfer more and more of the responsibility for fighting the Viet Cong and North Vietnamese Army back to the South Vietnamese forces. The US would continue to provide economic support, combat support (fire support, intelligence, and command and control), and combat-service support (logistics and supply) but would begin redeploying US combat forces out of Vietnam and focus on training the South Vietnamese to assume the entire fight in 1973.

Ensuring National Government Stability After US Counterinsurgency Operations

CHAPTER 1

INTRODUCTION

> Would anyone really be willing to fight and die for an entirely corrupt government and national military structure? Compare this to MacArthur in Japan. When he left Japan after running afoul of President Truman, Japanese citizens lined the streets with tears in their eyes because he had brought good governance. If our generals are not prepared to do this in conflicts like Iraq and Afghanistan, we should not get involved.
>
> –General Anthony Zinni[1]

Employing US combat formations to assist foreign states in killing their rebellious citizens during counterinsurgency (COIN) interventions should be considered a radical departure from normal state-to-state intercourse. This radical departure occurs when US combat formations are inserted into the internal conflict of another state and operate lethally and unilaterally to protect and win foreign populations and defeat foreign insurgents. In describing this style of US intervention in the Philippines in 1899, John Bass of *Harpers* called this a harsh and philanthropic kind of war.[2] US intervention of this stripe assumes that the host-nation government and security forces of another country cannot protect and win its own populace and defeat its own insurgents without US help. But what guarantees the survival of a US-supported state after the withdrawal of US combat formations?

What is the Problem?

In January 2014, I was a major teaching at the US Marine Corps' Expeditionary Warfare School in Quantico, Virginia, when I saw a news report that ISIS had captured Ramadi, Iraq. In January 2016, within days of my retirement ceremony from the Marines, I read about the Taliban regaining control of Marjah, Afghanistan. I had served as an infantry company commander in Ramadi, Iraq in 2008–2009 at the tail end of the surge and as an infantry battalion operations officer in Marjah, Afghanistan, in 2010–2011.

Numerous young officers, including myself, not only studied population-centric COIN theory and history, but we also taught it and practiced it for the better part of a decade and a half. We read from colonial counterinsurgents like Charles Gwynn, David Galula, Roger Trinquier, and Frank Kitson, as well as more contemporary synthesizers of classical COIN such as John Nagl, David Killcullen, David Petraeus, Andrew Krepinevich, James Mattis, and the US Army and Marine Corps Field Manual (FM) 3-24 *Counterinsurgency* (2006). Our battalions held weekly COIN seminars. Some of us wrote COIN articles in professional journals. We received language and cultural training before deploying into theater and benefitted from a largely supportive nation. We attended COIN academies in theater, conducted predeployment training at places like Twentynine Palms, California, with real Iraqi and Afghan role players and executed COIN operations with the best-educated American military force in US history. In comparison, Vietnam veterans, during my interviews with them, recalled serving as part of a conscript force and predeployment training consisting solely of boot camp and getting to fire a belt of fifty machine gun rounds as their preparation for Vietnam.

Introduction

In Ramadi, Iraq, at the tail end of the surge, US Marines were well respected by the Iraqi locals, sheiks, police, and the Iraqi Army. Marine Corps officers sat on almost all the local councils (e.g., the North Ramadi, South Ramadi, and East Ramadi Councils). We had won the trust, respect, and support of the populace. In Marjah, Afghanistan, battalions like 3d Battalion, 6th Marines, and 1st Battalion, 6th Marines paved the way into the heart of Helmand Province at substantial cost in American lives in early 2010. My battalion, 2d Battalion, 9th Marines (2/9), arrived in Marjah in the late summer of 2010 and began the deployment with between twenty-five to thirty violent significant acts (SIGACTS) per day. We operated alongside 2d Battalion, 6th Marines (2/6), and took over from the two previous battalions who had conducted the initial seizure of Marjah during Operation MOSHTARAK. After seven months, 2/9 had fifteen Marines and sailors killed in action, more than a hundred amputees or severely injured Marines and sailors, and one medal of honor recipient. However, in just one tour, 2/9 and 2/6 were able to reduce violent SIGACTS to nearly zero. Additionally, 2/9 had more than 200 Afghan students—both boys and girls—attending a single school in northern Marjah. We had won the support of the populace, and the Taliban could not operate effectively in Marjah while we were there.

I was at a loss. How was it possible to reconcile the fact that the US military could have been so successful at COIN while US forces were present and then have these gains unravel less than four years after US combat formations withdrew? Why—despite the differences in the US interventions in Iraq and South Vietnam, in terms of military leadership, professionalism of military forces, training, general support of the nation and massive technological advantage—were the outcomes in South Vietnam in 1975 and in Iraq in 2014 similar? As I wrestled with these questions, I began to ask, "Had the US ever succeeded in foreign COIN? If the US had, then when and where had this happened? Was it possible for US forces to succeed in COIN but fail to create a state that was able to govern and defend itself for longer than four years after US combat formations withdrew? Was it possible to create a US-supported state able to govern and defend itself after the US combat formations withdrew, even if US forces had not been successful in COIN operations?"

My questions were driven by more than academic interest. The US appetite for intervention in large-scale COIN may be diminished, but this has not lessened the number and threat of potential conflicts the US may be dragged into. The growing threats of brushfire wars all over the globe and possible interventions in places like Syria, Mali, Niger, and elsewhere presaged the need for a strategic methodology capable of producing US-supported states that are able to survive longer than four years after US forces withdraw. This would reduce the overall costs of large-scale COIN interventions to a fraction of their current costs, without increasing the average duration of these conflicts.

To put these questions into context and begin a search for possible answers, I analyzed the universe of US uses of force abroad from 1898 to the present. I identified a total of ten instances in which the US employed combat formations to intervene; and of these ten, nine instances where the US also led the foreign COIN effort.[3] The use of the qualifier—combat formations—is deliberate and provides an important distinction. Unilateral, direct combat operations where US combat formations operate as lethal agents killing rebellious citizens on behalf of a host-nation, is rare in the history of the US's use of force abroad. It constitutes de facto, if not de jure, appropriation of host-nation sovereignty. Intervention by US conventional combat forces provides a clear signal that the US policymakers perceive the host-nation's situation as being so dire that without a radical degree of intervention, the US-supported state will fail. According to Brent Gravatt, "Without a doubt, the surrender of the control of one's army, an executive agency, is concomitantly the surrender of one's sovereignty. Only a government experiencing a situation of extreme emergency will be willing to relinquish the command of its armed forces to a foreign power."[4]

In contrast, smaller-footprint COIN interventions do not necessarily constitute a major abdication of the international or material sovereignty of a state. Smaller-footprint COIN interventions might include the commitment of US special operations forces or the provision of economic, intelligence, or training support. These smaller-footprint support mechanisms constitute normal operations of US foreign policy abroad. They also reflect the relative confidence of US policymakers that, while fragile, the supported state is not in danger of imminent failure. This was the case in the US special operations support provided to the El Salvadorian government's COIN efforts in the 1980s and the Philippines since 2001.

MEASURING STATE SURVIVAL FOLLOWING US COIN INTERVENTION

What emerged from this analysis was that before 1950 American combat formations intervened twice as often in foreign counterinsurgencies but had durations just as long as US interventions after 1950. However, before 1950 the US-supported states tended to last an average of 35.7 years after the withdrawal of US forces, whereas after 1950 the US intervened half as often but produced states that have tended to last only 3.25 years (see table 1). What accounts for this dichotomy? What accounts for the fact that after 1950 the US intervened half as often, stayed just as long, but failed to produce states capable of surviving for more than 3.5 years without re-intervention with massive combat support?

To focus my subsequent research and analysis, I reframed these general questions into three more specific and measurable ones that are addressed in this book: Did the US-supported state survive after the US left? If it survived, how long did it survive? What accounted for one state's survival and another state's failure after US withdrawal? While questions of the quality of democracy produced by the intervention or achieving the stated objectives of US interventions are not unimportant, neither are they sufficient by themselves. For instance, it is pointless to have a discussion concerning the quality of democracy in South Vietnam in 2017 as a result of the US intervention there in 1965. This is because South Vietnam did not survive. Therefore, the most foundational measure of success for US interventions in support of the COIN operations of another state is whether the US-supported state survived and, if so, for how long.

To make sense of the data and to address these questions, I adopted two key measurements related to state survival after US intervention in large-scale COIN: duration and state longevity after withdrawal (SLAW). Duration is measured from the advent of unilateral combat operations by American formations to the withdrawal of these combat formations from combat or the country (see left side of table 1). State longevity after withdrawal (SLAW) is measured from the withdrawal of US combat formations from combat or the country until the violent overthrow of the US-supported state or a large portion of it (see left side of table 1). The use of the qualifier violent overthrow is deliberate. This is an acknowledgement that a potentially healthy outcome for an insurgency may be the reintegration of the insurgents back into peaceful political efforts of the state. The termination of the El Salvadoran insurgency in 1992 is a good example of this.

Table 1. US COIN Intervention Outcomes Before and After 1950.[5]

	Philippines	Panama	Nicaragua	Haiti	Dominican Republic	Cuba	USSR	Average Before 1950	Vietnam	Afghanistan	Iraq	Average After 1950
Combat Operations Commence	1898	1903	1927	1915	1916	1917	1918		1965	2001	2003	
Combat Operations Terminate	1913	1914	1933	1934	1924	1919	1920		1972	2014	2010	
Duration (years)	15	11	5	19	8	2	2	8.85	7	13	7	9
Violent Overthrow or Loss of Territory	--	1968	1979	1946	1965	1959	1920		1975	TBD	2014	
SLAW (years)	71/104[1]	54	46	12	41	26	0	35.7	3	TBD	3.5	3.25

Potential Reasons for the Dichotomy Before and After 1950

This book will compare contemporary theories for describing COIN success or failure compared to the degree of embeddedness used to develop host-nation security and governance institutions. These will be analyzed to identify potential correlations to increasing SLAW. The most common contemporary theories for articulating COIN success are population-centric COIN, enemy-centric or authoritarian COIN, troop-to-population ratios, interdiction of insurgent sanctuaries, and continued support to the host-nation government after withdrawal. Closely related to enemy-centric COIN theories are discussions related to how the strong lose in COIN and how the weak win. For this reason, I incorporate these discussions into my examination of enemy-centric COIN theories. I compile the evidence from these competing explanations in a table in chapter 6. This table will allow us to analyze the evidence latitudinally across all four cases to identify patterns and determine if correlations exist between these possible drivers and increases in SLAW.

For the purposes of identifying the impacts of population-centric COIN on SLAW, this book will use three criteria to establish evidence for analysis. The three criteria I will use are: presence, weight, and effectiveness. The three questions this book will ask of each case study is whether evidence exists to support the following questions: First, was a population-centric methodology present as part of the US COIN-intervention strategy? Second, if so, was the population-centric strategy weighted, or the principal focus of US COIN-intervention strategy for some portion of the intervention? Third, if such a methodology was present and weighted, was it effective? I considered the population-centric methods effective if they gained the support of the populace, controlled the movement of the populace, and protected the populace from insurgent control or reprisals.

Population-centric COIN relies heavily upon Mao Tse-tung's understanding of guerilla war and seeks to separate the insurgents from the support of the population. Population-centric COIN prioritizes the security and co-option of the population over defeating or suppressing the insurgency. This is not to say that population-centric COIN ignores the defeat of the insurgency. Instead, it deprioritizes defeat of the insurgency in comparison to securing and controlling the population. Much of the writing that population-centric COIN theory is based on evolved from colonial counterinsurgents like Trinquier,[6] Galula,[7] Thompson,[8] Kitson,[9] and Gwynn.[10] What made them unique compared to modern counterinsurgents is that as colonial counterinsurgents, they were not merely supporting the host-nation government, they were the host-nation government. As such, they were not influencing host-nation institutions from the outside, but they were occupying and inhabiting them from within.

Because of the US's failure in Vietnam, large-scale COIN was eschewed from 1975 to 2004 as being inconsistent with the US way of war. The sentiment seemed to be justified by the success of the Powell-Weinberger Doctrine in the Persian Gulf War which said the US should only go to war when it was prepared to use overwhelming force and when it had a clear US interest involved and the support of the US population.[11] With the Afghanistan and Iraq wars though, a new generation of scholars and scholar-practitioners began to leverage the writings of the colonial counterinsurgents in updating population-centric COIN theory. They argued that the central requirement was not only the security and co-option of the populace but also the removal of the underlying causes of dysfunction that had caused the insurgency in the first place—the removal of causational factors of instability.[12]

As a result of these scholar-practitioners' efforts, the 2006 and 2014 versions of the DoD's COIN manuals added the requirements for US personnel to enhance good governance, rule of law, and provision of essential services, in addition to securing and co-opting the populace.[13] Therefore, while US forces pursued insurgents kinetically and secured the populace, they were also responsible for removing the causational factors allegedly responsible for the instability in the first place.[14]

I will use the same three criteria and questions I used with population-centric COIN to identify the impacts of enemy-centric or authoritarian COIN on SLAW. With regard to a measure of historical effectiveness, I considered enemy-centric methods effective if they defeated, destroyed, or removed the insurgency as a legitimate and lethal challenger to the US-supported state.

Enemy-centric or authoritarian COIN focuses the counterinsurgent's efforts on suppressing or defeating the insurgents rather than protecting the populace or addressing the grievances that caused the insurgency. The potential efficacy of enemy-centric or authoritarian COIN has been reinvigorated with the writings of Gil Merom,[15] Martin Van Creveld,[16] David Ucko,[17] and Daniel Byman.[18] Merom argued that insurgents are able to use the sensibilities of Western liberal democracies against them. He assessed that modern democracies have minimal tolerance for protracted campaigns and high casualty rates and are repulsed by suicide bombings which gives insurgents an advantage against them. Merom pointed out that this advantage does not exist when insurgents battle authoritarian states that are less politically affected by suicide bombings, high casualty rates, and collateral damage. Van Creveld articulated what he termed the Hama Model used by Hafez Al Assad to crush the Muslim Brotherhood in Hama, Syria, in 1982. Ucko also provides historic reviews of authoritarian-Marxist COIN tactics employed effectively, if not ruthlessly, against societal reactionaries. Additionally, Byman argues, "Authoritarian states are often surprisingly successful counterinsurgents. In particular, authoritarians often repress on a vast scale and inhibit insurgent organization, transfer populations, have excellent intelligence penetration, and can counter war weariness in ways not available to democracies."[19] As a result, enemy-centric or authoritarian COIN has once again gained some traction. As David Ucko points out, authoritarian COIN has been enormously successful at suppressing the grievers, rather than addressing the grievances, which is the basis for contemporary population-centric COIN.[20]

In many cases, after their own successful revolutions, many former revolutionaries like Fidel Castro, Ho Chi Minh, Joseph Stalin, and Mao leveraged very effective enemy-centric/authoritarian COIN strategies and crushed social reactionaries or alleged counter-revolutionaries. Stalin's Kulak Revolution, Mao's Cultural Revolution, Ho's counter-revolutionary campaign, and Castro's suppression of reactionaries have demonstrated the potential effectiveness of an enemy-centric/authoritarian COIN campaign, albeit generally requiring large-scale massacres to achieve their ends.[21] Ucko however, also points out that whereas these may present successful short-term strategies for authoritarian regimes, they are neither tenable for western liberal democracies politically nor demonstrate great potential for long-term success. Ucko demonstrates that as soon as the authoritarian suppression is lifted, the insurgency and its grievances explode even more dynamically, as has been observed in Chechnya, the former Yugoslavia, and Syria for example. Still, none of these writers on enemy-centric/authoritarian COIN argue for how these methods assist or prevent a third-party counterinsurgent in leaving behind capable host-nation institutions after withdrawal.

A tangential aspect of enemy-centric COIN is a body of knowledge that seeks to address how the strong lose and the weak win in COIN. Andrew Mack looked at dichotomies in why insurgencies have stronger motives to win and argued that for the intervening counterinsurgent, the fight is a war of convenience where the fight for the insurgent is an existential conflict for survival. He noted that this limited the degree to which the intervening counterinsurgent was willing to sacrifice relative to the insurgent, which helps explain why intervening counterinsurgents are more sensitive to casualties and protracted fights than the insurgents.[22] Ivan Arreguin-Toft examined how mismatches in tactical methods may give insurgents an advantage. He noted that when counterinsurgents attempted to employ direct conventional methods to defeat insurgents who were employing an indirect unconventional defense, this gave the insurgent an advantage.[23] As mentioned previously, Merom noted how insurgents successfully use Western liberal sensibilities against them,[24] and Jeffrey Record argued that external support for an insurgency is a necessity for success and can minimize or negate altogether the material advantage of a counterinsurgent.[25]

These theories have a few gaps as to how they relate to third-party interventions and SLAW. First, they seem to approach these conflicts as static rather than dynamic enterprises. They assume that intervening third-party COIN force and the insurgents began with and ended with the same tactical approaches (i.e., direct conventional versus indirect unconventional). The reality is—in the case the authors mentioned most often the US war in Vietnam—both the US and the insurgents adapted constantly and switched between direct and indirect methods for over nine years. What these theoretical approaches ignore is a glaring concern —the only entity that was not required to constantly adapt was the element the US left behind, the host-nation. Second, the authors, though they do not expressly indicate, nevertheless assume that the crucial fight that is to be won or lost is between the intervening third-party COIN force (e.g., American, British, French) and the insurgents. This leads to the conclusion that the reason that third-parties lose the will to continue to fight is that they are fighting a war of convenience whereas it is a war of survival for the insurgent. Again, this ignores the glaringly obvious —if this is a war of survival for the insurgency and it drives them to levels of sacrifice that the third-party is unwilling to pay, what about the host-nation government on whose behalf the third-party intervened? Certainly, theirs is also an existential fight every bit as great as the insurgents'? So, whereas their theory can explain why the US would eventually want to look for a way out of an expensive and bloody foreign COIN fight, it says little about why the host nation does not compete with the same intensity as the insurgents. Third, the authors look at how insurgents exploit the sensitivities of Western liberal democracies but are unable to do the same with autocratic regimes. As previously mentioned, enemy-centric/authoritarian methods have been effective; however, what is missing here is that the US was often successful at both enemy-centric and population-centric COIN methods while they were present and yet, after 1950 they still failed to produce regimes that lasted more than 3.5 years when they withdrew unless they re-intervened.

The difficulty in analyzing American foreign COIN campaigns from a perspective of strictly population-centric or enemy-centric methods is that this is, for the most part, a false dichotomy. This book does extract evidence of population-centric and enemy-centric COIN methods separately for analysis, but this is not because they were instituted and executed in a mutually exclusive fashion. Rather, in most cases, the US has tended to either disproportionately focus on the population or the insurgents at different times, but not to the degree that it completely ignored either. This is important as this book does not make an argument in favor of either method but merely acknowledges what most hybrid-COIN theorists argue.

Hybrid or situational COIN is nothing new. This COIN method balances the rights and responsibilities of the counterinsurgent and the population. It originates from Francis Liber's *Guerilla Parties* (1862), which was ultimately codified into *General Order 100* (G.O. 100) issued by Abraham Lincoln on April 24, 1863.[26] General Order 100 prescribed the obligations of the counterinsurgents to protect the population and their property. It also enumerated the responsibilities of the population to respect the counterinsurgent's authority and prescribed penalties for refusal to do so. This was expressed as Lincoln's policy of moderation and reprisal. It was later reformulated in the US's first foreign COIN intervention in the Philippines as attraction and chastisement.[27] It was further reprised in Secretary of Defense Mattis' dictum to the US Marines in Iraq as "No better friend; No worse enemy." Indeed, the whole of US COIN history can be argued to have ostensibly used a series of hybrid-COIN methods. Therefore, this book acknowledges the hybridity inherent in any US foreign COIN campaign. It also examines cases longitudinally by

Introduction 13

separating population-centric and enemy-centric evidences to discern any patterns which discretely produce increases or decreases in SLAW.

Identifying the optimal troop ratios for COIN operations is a commonly cited variable in COIN success or failure. The 2006 US field manual on COIN estimates that "most density recommendations fall within a range of 20 to 25 counterinsurgents for every 1,000 residents in an [area of operations]. Twenty counterinsurgents per 1,000 residents is often considered the minimum troop density required for effective [counterinsurgency] operations..."[28] This translates to one counterinsurgent for every fifty citizens. Steven Goode noted that this was most likely reflective of James Quinlivan's seminal work on the subject.[29] John McGrath argued that mitigating factors like intensity of the conflict, densities of population, troop rotation, and use of indigenous forces would affect optimal troop densities but neither he nor Goode seriously contradicted Quinlivan.[30] Rather, discussions of appropriate COIN troop ratios reflect mere adjustments to Quinlivan's measurements, or they question his variables.[31] McGrath added to the discussion by pointing out the multiplicative impact of indigenous forces as a replacement for or an amplifying effect to US troop ratios. As a result, the minimum number required might possibly be modified from 1:50 to possibly as much as 1:91 counterinsurgents to citizens.[32]

To determine an adequate range of troop-to-population ratios, I set the most conservative estimate of this range as one counterinsurgent for every fifty citizens (1:50) as identified by the DoD's COIN manual. For the most liberal estimate of this range, this book establishes this as one counterinsurgent required for as many as ninety-one citizens (1:91). Therefore, this book will rely on this range, 1:50 to 1:91 counterinsurgents per citizen as an adequate troop-to-population ratio and will use this to determine if a correlation exists between adequate troop ratios in COIN and increases in SLAW.

The inability of the US and the supported state to foreclose on insurgent sanctuaries has been a common theme associated with predicting COIN success or failure. In this book I will determine if there is any correlation between the ability of the US to foreclose on insurgent sanctuaries and SLAW. To establish whether the US foreclosed on the use of insurgent sanctuaries, there needs to be evidence that the insurgents were not just chased out of their sanctuaries temporarily. Instead, there needs to be evidence that the US permanently prevented insurgent use of these sanctuaries and foreclosed on most, if not all, insurgent controlled areas.

Harry Summers was the second to last US officer out of Saigon and was in the last helicopter that lifted off from the US Embassy in 1975. He related in his seminal analysis of the conflict that it was not insurgents who captured Saigon and defeated the South Vietnamese government, but rather North Vietnamese infantry supported by tanks and artillery. Summers argued the US did not lose a counterinsurgency as the Viet Cong (VC), the insurgent force the US was fighting, had already been defeated in 1968 and were an inconsequential battlefield entity after that. He argued that US forces would have been better employed in creating a broader demilitarized zone (DMZ) and focusing US efforts on defeating the North Vietnamese Army (NVA) while allowing the South Vietnamese forces to prevent the return of the VC to South Vietnamese hamlets.[33] Marine Corps General Anthony Zinni and Max Boot also agree that winning a counterinsurgency is improbable where insurgent sanctuaries are never successfully interdicted.[34]

The continued provision of military or economic support after the withdrawal of US combat formations has been another common theme associated with predicting COIN success or failure. I attempt to determine if there is any correlation between the continued provision of US military or economic aid after the withdrawal of US combat formations and increases to SLAW. I also look at the provision of continued US combat support and combat-service support after the withdrawal of US forces. Combat support includes artillery, tactical air support, assault support, engineering, intelligence, and communications support. Combat-service support is related to the provision of sustaining functions such as logistics, supply, maintenance, and transportation. This book will therefore look to identify any correlation between the continued provision of US combat support or US combat-service support and SLAW.

Many of the arguments for continued US support as a defining factor in increasing SLAW in Vietnam are related to the Foreign Assistance Act of 1974. This act contained an amendment prohibiting further US combat support to the South Vietnamese government and military. General Zinni and Max Boot argue that, based on the performance the South Vietnamese military during the 1972 Easter Offensive, the South Vietnamese could have defeated the North Vietnamese in 1975 if they continued to receive US combat support, combat-service support, and funding at levels they had previously received in 1972.[35] Generals Raymond Odierno, James Amos, Joseph Dunford, Jack Keane, James Mattis, Lloyd Austin III, David Petraeus, and former Secretary of Defense Leon Panetta have also made similar cases that the absence of US advisors, combat support, and combat-service support after 2010 contributed to ISIS's success and the Iraqi military's failure in 2013 and 2014.[36] Therefore, this book will attempt to confirm or deny any relationship between continued US combat support or combat-service support and SLAW.

Contemporary COIN theories have great value in understanding tactical approaches to COIN. However, the US has not typically intervened or assumed the lead for a COIN effort because it sought to protect a foreign populace or to defeat a foreign insurgency. Rather, the US has intervened or assumed the lead in a foreign COIN fight when the host nation could not secure and control its own population nor defeat its own insurgency. Had the host nation been able to do either of these or both, the US would not have expended US lives and money to do so. As such, contemporary COIN theories are particularly unsatisfying when trying to understand what allows a US-supported state to endure after the US has withdrawn. In contrast, scholarship related to trusteeships and neo-trusteeships, military government, shared sovereignty arrangements, armed state building, and military adaptation provide unconventional but potentially more gratifying possibilities.

While all US COIN interventions share many tactical similarities, those before and after 1950 differ fundamentally in the degrees of embeddedness the US used to develop host-nation institutions. Degree of embeddedness here refers to how deeply US officers were embedded within and inhabited host-nation institutions to remediate or replace "government which is unstable, inadequate, or unsatisfactory for the preservation of life..."[37] Degree of embeddedness also refers to the amount of host-nation sovereignty abrogated by the host-nation and appropriated by US civilian and/or military officers.

Prior to 1950, US military and diplomatic forces addressed state failure by relying on strategies of high embeddedness in the course of COIN interventions. US officers repaired or replaced host-nation institutions from inside these institutions by inhabiting them. These highly embedded strategies relied on a more contingent view of host-nation sovereignty and may have been, as Richard Caplan concedes, the "least worst option".[38] After 1950, US military and diplomatic forces addressed state failure by relying on strategies of low embeddedness in the course of COIN interventions. US officers repaired or replaced host-nation institutions from outside the host-nation institutions by influencing them.

State-centric COIN Theory

My core argument in this book is that the degree of embeddedness used to develop the host-nation governance and security forces during a third-party intervention is a more effective predictor of SLAW increases than contemporary COIN theories. Specifically, I argue that SLAW improves in the course of COIN interventions when US officers deeply embed within and inhabit host-nation institutions (institution-inhabiting strategies). Inversely, when the US employs strategies of lower embeddedness (institution-influencing strategies), then host-nation SLAW decreases in the course of COIN interventions. Furthermore, I argue that symbiotic adaptation is the causal mechanism that accounts for the greater SLAWs associated with institution-inhabiting strategies. There are two antecedent conditions under which my theory would be valid. The first condition is a perception that state failure or erasure is either imminent or already in effect. The second condition is the perception that only the intervention of US combat formations is sufficient to arrest the instability caused by actual or imminent state failure.

Institution-inhabiting strategies historically have consisted of trusteeships, military governance, shared sovereignty arrangements, encadrement, and the like. Trusteeships, military government, and encadrement reflect the deepest penetration and inhabitation of a host-nation's institutions. A trusteeship exists when foreign military and/or political officers occupy legal, political, police, or military offices and exercise authority normally reserved solely for host-nation officers. Possession of these offices allows US military and civilian officers to make and enforce laws, establish and collect taxes, run elections, arrest, and use lethal force to defeat an insurgency.[39]

Military governance is a specific form of trusteeship wherein foreign military officers assume political roles in the host-nation's government. They govern until such a time as they can be replaced by civilian authority, either from their own country or representatives from the host-nation.[40] Before 1950, military governance reflected the predominant means of developing states during small wars or after the conclusion of large ones like World War I & II.[41]

Shared sovereignty arrangements are a more circumscribed version of trusteeships. They are relationships wherein foreign parties occupy specific roles within the offices of certain host-nation institutions. An example of shared sovereignty might be a customs receivership. Customs receiverships exist when a foreign party installs its own officers in the customs institutions of another state to prevent corruption, ensure accountability, and usually collect an international debt.[42] Shared sovereignty arrangements are more circumscribed than trusteeships in the degree to which they appropriate the host-nation's sovereignty. The foreign officers operating within state organs may run specific parts of the host-nation government, or, as happened in Sierra Leone, they may simply have a veto authority as a condition for continued foreign support.[43]

Finally, encadrement may be a facet of trusteeship or shared sovereignty arrangement. Encadrement is a term appropriated from Bernard B. Fall that describes the command of a host-nation's security forces by foreign military officers.[44] These foreign officers exercise the same legal command authority as host-nation officers including the power to command troops in the field, reward, punish, promote, and demote. These foreign officers may be authorized and paid for through legislation by the foreign party alone, by the host-nation government, or by both.[45] The host-nation security force may be the only part of the host-nation government that foreign officers occupy (shared sovereignty arrangement). Or the host-nation security force may be only one among many of the institutions directly run by foreign officers (trusteeship). Encadrement was the default method employed by the US military before World War II to repair or replace host-nation security forces during small wars interventions.

In contrast, institution-influencing strategies have been the default strategies the US has employed to develop host-nation governments and security forces during COIN interventions since May 1955. Institution-influencing strategies refer to foreign political and military officers advising host-nation governments and/or security forces. Influencing strategies lack any legal command authority or legislative, executive, or judicial power to force change. They rely entirely on influence to bring about change in the host-nation government and its security forces. Since May 1955, all of the US military's efforts to develop or improve host-nation government and security force institutions have taken place outside these institutions through influencing operations.

Examples of influencing strategies include governance assistance, conditionality agreements, military assistance to governance, and advise and assist operations. Governance assistance reflects foreign civilian political assistance rendered to a host-nation government. An example of this includes the use of ministerial assistance teams (MATs) in Iraq that advised Iraqi ministries in 2006.[46] Conditionality agreements are tools generally associated with neoliberal investment in developing countries. These agreements provide continued investment on condition of the performance of the developing state in accordance with established requirements of a foreign party. The International Monetary Fund (IMF), as the lender of last resort, is commonly associated with conditionality agreements. A form of military conditionality agreement exists when continued economic or combat support by a foreign military is contingent upon specified actions taken by the host-nation security force.

Military assistance to governance is an influencing strategy that consists of military officers advising civilian host-nation government officers. The purpose of military assistance to governance is to develop or refine basic governance capacity in the areas of rule or law, elections, and provision of essential services.[47] Advise and assist is the military operational counterpart to governance assistance, except that military advisers work with host-nation security forces and advise them in how to further develop security capacity. Security cooperation is the least embedded form of influencing operation and allows both the foreign military and the host-nation security force to train together to develop the ability of both militaries to integrate and work together.

However, in all these influencing strategies, where the host-nation is reluctant to accept advice and to reform, there is little that US officers can do to compel change. The only compelling mechanism a foreign military possesses is coercive influence in withholding money or support when the host-nation is reluctant to change. This coercive influence seeks to force a host-nation government or host-nation security force to behave a certain way. Nevertheless, even in the case of poor, self-destructive, or even illegal/immoral performance by the host-nation security force, the foreign officers have no way to command and compel change.[48]

SLAW Outcomes Associated with Degrees of Embeddedness

US COIN interventions before 1950 produced longer lasting states than interventions after 1950 (see figure 1). Where the degree of embeddedness was highest, states tended to persist ten times longer on average.

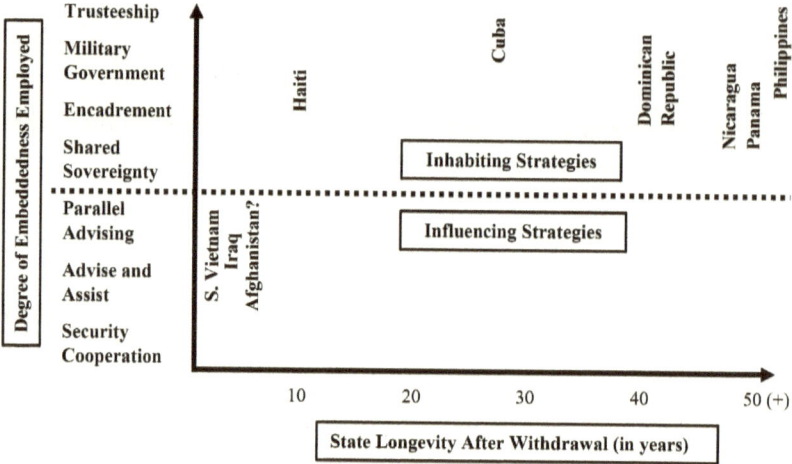

Figure 1. State Longevity Outcomes: Inhabiting Versus Influencing.

Figure 1 depicts the variegated SLAW outcomes resulting from the use of different degrees of embeddedness to develop host-nation government and security forces during US COIN interventions. The vertical axis denotes the degree of embeddedness employed with the most invasive strategies at the top and the least invasive strategies at the bottom. The horizontal axis denotes the SLAW in years. The dotted line demarcates a theoretical separation between inhabiting strategies on the top and influencing strategies on the bottom. The results are dramatic. Where inhabiting strategies were used, SLAW increased dramatically. Where influencing strategies were employed, SLAW was limited precipitously. Thus, there appears to be a correlation between how deeply US military and diplomatic forces embedded in host-nation institutions and variations in state longevity. Interestingly, SLAW was not the only outcome influenced by the degree of embeddedness used by the US. The governance and security outcomes related to the use of different degrees of embeddedness were also significant.

Figure 2. Governance and Security Outcomes Related to Degree of Embeddedness.

		Degree of Embeddedness-Governance	
		Higher-Institution Inhabiting	**Lower-Institution Influencing**
Degree of Embeddedness-Security Forces	**Higher-Institution Inhabiting**	**Stumbling State** High Security Low Democracy	**Tumbling State** High Security Faux Democracy
	Lower-Institution Influencing	Not Observed In This Research	**Strategic Rentier State** (US combat support) **Crumbling State** (loss of combat support) Large areas of No Security No Democracy

Whereas the previous table depicted state outcomes with respect to SLAW, figure 2 depicts outcomes of COIN intervention strategies in developing host-nation governance and security. Horizontally, across the top of the figure, the degree of embeddedness used to develop host-nation governance is depicted, and vertically, along the left-hand side, the degree of embeddedness used to develop host-nation security is depicted. Both along the top and on the left-hand side, higher embedded strategies relate to inhabiting strategies, and lower embedded strategies relate to influencing strategies.

Combinations of different degrees of embeddedness and their impacts on security and governance produced three types of states after US withdrawal: stumbling states, tumbling states, and crumbling states. Whereas the names of these different types of states are not intended to be overly pessimistic, it is important to recognize how broken a state must be before the US will intervene with combat formations to save it. I also include an analysis of the quality of governance and democracy the intervention created as an additional measure of effectiveness of liberalizing interventions.[49]

Quadrant #1 (upper left) depicts where inhabiting strategies were used to develop both host-nation governance and security. The result of this combination of highly embedded strategies in governance and security development produced a stumbling state, or a state defined by high security and low democracy. I define low democracy as a situation where democratic institutions persist indefinitely but are restrained by systemic corruption associated with patronage systems such as the Filipino *compadrazgo* system.[50] A low democracy appears to be the zenith of what might be expected as a result of third-party COIN intervention with combat formations—at least for some significant time after the withdrawal of US forces. Whereas a best case of low democracy may seem unduly pessimistic, Paul Miller also observes that, "Few state-building missions result in an unqualified and comprehensive success."[51] The idea of a stumbling state conjures up the image of a state recovering from frailty but nevertheless providing semi-effective and persistent governance, albeit beset by routine but nonexistential shortfalls. The Philippines provide a good example of a stumbling state.

Quadrant #2 (upper right) depicts the results where a combination of inhabiting and influencing strategies were used to develop host-nation security forces and governance respectively. This combination of low embeddedness to develop host-nation governance but high embeddedness to develop host-nation security forces produced tumbling states. Tumbling states are exemplified by high security and faux democracy. That is, these states possess very effective security instruments but poor governance and/or democracy. Faux democracy describes a reality where only the pretense of democracy exists. This pretense of democracy is maintained by an authoritarian regime that consistently abrogates old laws and fabricates new ones to justify what is currently illegal or democratically corrupt. This combination of high security and faux democracy points to a lengthy authoritarian reign. It also prognosticates an inexorable demise when the state's security force(s) becomes less effective or government ineptitude or corruption are no longer tolerable by the populace. Nicaragua (1933–1979) is a solid example of a tumbling state.

Quadrant #3 (lower right) denotes the use of strategies of low embeddedness to develop both host-nation governance and security. In the short-term, the use of strictly influencing strategies produced strategic rentier states. Hossein Mahdavy and Hazem Beblawi provide the first identifications of what constitutes a basic rentier state.[52] A rentier state is one in which foreign rents constitute a significant portion of its economic capacity and for whom there is a conspicuous absence of domestic revenue from taxation.[53] According to Beblawi, there are four criteria that describe a rentier state: 1) rent situations predominate, 2) the state economy relies in large part on foreign rents, 3) only a small percentage of the population in the rentier state is involved in the production of the rent, even if many benefit from it, and 4) the host-nation government is the primary recipient of the rents.[54] Because foreign rents replace many of the needs for domestic tax extraction from the populace, there is a twofold stagnating effect. First, massive welfare regimes create little incentive for citizens to pressure the host-nation government for reforms. Second, the lack of popular pressure increases the likelihood of kleptocratic regimes and decreases the likelihood of change.[55]

Barnett Rubin and Rolf Schwarz extended the concept of foreign rents and conceptualized the idea of strategic rents. They applied the concept of foreign rents for natural resources like oil to the extraction of foreign rents in the form of foreign aid for military basing agreements, alliances, and other similar arrangements.[56] This book extends their work to describe the production of strategic rentier states resulting from institution-influencing strategies.

When states receive strategic rents in increasing amounts based on a deteriorating security situation, a problem of moral hazard evolves. This produces what Beblawi terms a break in the reward-causation cycle, which creates what he calls the "rentier mentality".[57] In this case, there is greater incentive for the host-nation to perpetuate its own instability—the worse the security situation, the more the US is willing to pay to repair it. Krepinevich describes this type of situation in US efforts to develop the South Vietnamese security forces, "This outpouring of aid from the United States appeared to have an adverse effect upon the fortunes of the South Vietnamese. The more assistance that was given, it seemed, the less hopeful the prospects of defeating the insurgents became."[58] Because this perpetual dysfunctionality threatens US prestige in countries where the US has taken responsibility for security outcomes, the threat of continued dysfunction guarantees the payment of continued rents—at least until American political will or economic capacity evaporate.

This also is the case with states that result from institution-influencing strategies. As long as strategic rents in the form of combat support, economic support, or combat-service support continue to be provided, the state survives. However, once the strategic rents upon which the host-nation has become dependent are withheld, then the state becomes a crumbling state. A crumbling state is indicated by a state where all or vast parts of its territory have been overrun, and therefore this territory possesses no state provided security nor democracy. The overthrow of the entire regime, or large portions of the state's territory, make discussions of other possible measures of intervention success, like the quality of democracy, moot where the state no longer exists. Examples of this include South Vietnam and a large part of Iraq from 2013 and until US forces re-intervened.

How the different parties involved in an insurgency-counterinsurgency evolve and adapt through the conflict helps describe how a third party contributes to produce a stumbling, tumbling, or crumbling state. John Nagl and Theo Farrell provide excellent studies in how counterinsurgents adapt or fail to adapt in foreign COIN interventions. Both identify the qualities that allow or restrain foreign counterinsurgents from effectively adapting in COIN.[59] However, both miss a crucial point in foreign COIN intervention—the adaptation of the host-nation forces who are not leaving.

Carl von Clausewitz, in his seminal text *On War,* described war as a wrestling match between two hostile irreconcilable parties both trying to impose their wills on the other or *zeikampf,* or "two war." Through this contest of wills, each force is compelled to continually act, observe, react, and adapt to the efforts of the adversary.[60] This concept of mutual adaptation was further developed by John Boyd. Boyd's formulation of what has become known as the Observe, Orient, Decide, Act Loop (OODA Loop).[61] I describe these two processes as mutual competitive adaptation and they have serval implications that were originally predicted by Boyd in *Destruction and Creation.*[62] The first is that because of adaptation, the state of affairs that existed at the beginning of the conflict would bear little resemblance to the state of affairs at the end. Second, the combatants themselves would also be dramatically changed by the struggle as both are compelled to continuously learn and adapt until one can no longer keep up and either quits or is destroyed. This describes two actors well but does not account for the three or more actors involved in third-party COIN: the insurgent force, the US forces, and the host-nation forces.

This book describes artificial adaptation as a condition where the US artificially transfers uniquely American adaptations and lessons to the host nation, which had no part in the development of these adaptations and lessons and lacks the economic and cultural capacity to sustain them after the US withdraws. When the US operates outside of the host-nation's institutions, unilaterally pursuing the enemy, it often replaces the host-nation forces and takes the lead in pursuing the insurgents, shielding the host-nation's forces from destruction. US forces and the insurgents are forced to constantly act, react, and adapt to each other's actions and adaptations. Consequently, the insurgents who survive are trained by the experience. The US also learns its own lessons, but these are based solely on US military culture and technological and economic capacity. In contrast, the host-nation's forces are shielded from the need to adapt; their growth is stunted and they inherit uniquely American lessons—artificial adaptations—and lack the martial culture and economic and technological capacity to sustain these without perpetual US support. These states become strategic rentier states.

In contrast, symbiotic adaptation exists when US forces are constrained to adapt from within host-nation institutions and their discrete cultural and economic realities and employ institution-inhabiting strategies. The American officers are forced into mutual competitive adaptation with the insurgents, but the American officers compete from within the host-nation's institutions. The outcome of this symbiotic relationship is neither wholly American nor host nation; both are changed. In addition, because US officers are compelled to negotiate with the populace and compete with the insurgents through the host-nation institutions, all adaptations and lessons become the property of the host-nation institutions. Further, these lessons and adaptations are uniquely suited to the host-nation's discrete cultural dimensions and threats. The institutional adaptations that result from the symbiosis between US forces and host-nation institutions are solidified and become permanent through what Joel Migdal describes as a process of compliance, legitimation, and participation with the new institutions.⁶³

TESTING STATE-CENTRIC COIN

I made a couple critical choices when I developed and tested a theory of state-centric COIN. I chose to use the US as the central actor in all the cases, not because the US has had the best record in third-party COIN but to control for the disparate potential cultures of possible counterinsurgents from European, to Asian to African militaries. This was a way to control, as much as one can, for differences in martial culture, economic capacity, and systems of governance. Keeping the same actor throughout all the cases allowed me to clearly identify what stayed the same and what changed in terms of drivers for increases to SLAW. I also focused on cases where the US intervened with conventional combat formations because this is indicative of just how dire the US policymakers saw the situation. Had the situation not been so dire, US policymakers would have most likely intervened with a smaller footprint such as Special Forces and other forms of support. Lastly, I further narrowed my research to cases where the US was in the lead for the COIN fight. I did this to best isolate who was really making the tactical decisions and strategy. The one case in which the US intervened with combat formations but was not in the lead was the US COIN efforts as a subordinate part of a coalition in the Soviet Union in 1918.

I also do not go into detail about the tactics and strategies employed by the insurgents. This is also deliberate. I did not do this because the insurgent is unimportant. In fact, the opposite is true—the insurgent is critical. However, this book stipulates that the very presence of US conventional forces being necessary to defeat a foreign insurgency is indicative of the potency of the insurgency. This book further lays down that the tactics used by the insurgents greatly affect the reactions by the US as a third party to the COIN fight. However, what is often missed in examinations of third-party COIN success or failure is the most critical element—the host nation which will remain after the US leaves. The question then really is not whether the US and the insurgents effectively adapted in response to each other, but whether the US prepared the host nation to effectively adapt and respond to the insurgents after the US withdrew.

I employ small N-qualitative case study methodologies to both develop and to test a theory of state-centric COIN. These methods include the use of process tracing, contextually constrained historical comparison, Delphi method, and hoop tests to infer a theory of state-centric COIN and to test it alongside rival competing theories. This book utilizes a contextually constrained historical comparison of the universe of US COIN interventions involving major combat formations to determine antecedent conditions under which my theory would operate and identify what elements are responsible for increasing and retarding SLAW. Case study methodologies are well suited to testing predictions that are unique and not otherwise predicted by any other theory, testing of explanatory theories, and adding distinct explanatory value.[64]

My selected case studies provide fertile ground for theory development and testing.[65] A large N-qualitative study would not be required here as the entire universe of US interventions with combat formations affords only ten instances in total. Additionally, because the US military is the recurring actor in each of these cases, the inferences provided by my theory provide not only longitudinal depth within each case but also latitudinal comparison across all the cases in order to arrive at these unique predictions.[66]

The entire population of US COIN interventions from 1789 to the present amounts to twenty-two cases. From these, there were fifteen wherein the US military was the lead intervening military force. From these fifteen, there were a total of nine wherein the US military force was in the lead for the COIN intervention and conventional forces were employed. These cases are the Philippines (1898–1913), Panama (1903–

1914), Haiti (1915–1934), Dominican Republic (1916–1924), Cuba (1917–1933), Nicaragua (1927–1933), Vietnam (1965–1973), Afghanistan (2001–present), and Iraq (2003–2010).[67]

In order to test my theory against other possible explanations, I selected cases based on method of difference, extreme variance on possible drivers for SLAW and outcome with respect to SLAW, cases of most representativeness, and the host-nation government status at the onset of the US intervention. I selected two cases to represent US interventions before 1950—the Philippines and Nicaragua. The Philippines case demonstrates a comprehensive institution inhabiting case study as the US utilized an entirely trustee form of relationship with the host-nation as well as the encadrement of its host-nation security forces. This intervention represents the most encompassing appropriation of host-nation sovereignty but also the longest state longevity—a minimum of 71 years (see table 1). The Nicaragua case study demonstrates a unique combination of inhabiting and influencing strategies that nevertheless produced a regime longevity of 46 years (see table 1). This allowed me to distinguish potential outcomes when different degrees of embeddedness were used to develop governance and security in the same intervention. It also provided a case where the US failed in COIN efforts but succeeded nonetheless in developing a very effective host-nation security force.

I selected two more cases to represent the period after 1950—South Vietnam and Iraq. Vietnam exemplifies the institution-influencing phenomena through military assistance to government and advising and produced a regime longevity of only three years. Vietnam was also the first time the US intervened in large-scale COIN without using highly embedded strategies. Iraq also exemplifies the institution influencing phenomena through military assistance to government and advising and produced a regime longevity of only 3.5 years (see table 1).

This book will first pair case studies by instances in which the US intervened in tabula rasa governance conditions or created these conditions as a result of US intervention. I have paired a case of high embeddedness with a case of low embeddedness intervening in tabula rasa governance of the host nation. These two cases are the Philippines and Iraq. I have further paired a case of high embeddedness with a case of low embeddedness in which the US intervened in existing host-nation governance. The Nicaragua and Vietnam case studies represent these historically constrained contextual comparisons. In the final chapter, I compare all these case studies latitudinally against all the others to test my theory and identify correlations between different possible COIN theories and increases in SLAW.

Roadmap for The Book

Chapters 2 and 3 will examine cases of high embeddedness and low embeddedness in interventions where governance had been completely erased—tabula rasa (The Philippines, 1898–1913, and Iraq, 2003–2010). The Philippines represents the case of high embeddedness, and Iraq the case of low embeddedness. These cases will determine if there is a correlation between degree of embeddedness and state longevity in the course of US third-party COIN interventions where the state has completely failed or been erased. In contrast, chapters 4 and 5 will examine cases of high embeddedness and low embeddedness in interventions where the state was not yet failed but was on the verge of failure when the US intervened (Nicaragua, 1927–1933, and Vietnam, 1965–1972). Nicaragua represents the case of high embeddedness, and Vietnam the case of low embeddedness. These cases will determine if there is a correlation between degree of embeddedness and state longevity in the course of US third-party COIN interventions where the state has not completely failed yet or been erased.

Each of the case study chapters, chapters 2–5, will follow a similar pattern of presentation. Each will begin with a historical overview of the conflict. Next, the potential drivers for increasing SLAW during COIN interventions will be presented in COIN, troop ratio, insurgent sanctuary interdiction, and continued support after US withdrawal sections. Then, each chapter will look how the US developed governance and security in the case and the degree of embeddedness used. The chapters will then examine the outcomes of each state's initial tests after the withdrawal of US forces and then will wrap up the evidence in each case in the conclusions section.

The concluding chapter assembles the evidence from all the case studies to determine which possible explanations correlate most closely with SLAW. This chapter will then identify the general trends in the use of different degrees of embeddedness in SLAW, security, and governance outcomes. It will also examine how tenable high embedded strategies may be today and recommend possible policy implementation guidelines for their employment in the future.

Notes

1. Author's interview with General Zinni, January 2017.
2. Stanley Karnow, *In Our Image: America's Empire in the Philippines* (New York: Ballantine Books, 1989), 155.
3. The US used combat formations in the USSR as part of a coalition but US forces were not in the lead.
4. Brett L. Gravatt, *The Marines and the Guardia Nacional de Nicaragua 1927–1932* (Durham, North Carolina: Duke University, 1973).
5. 1-US combat formations ceased unilateral combat operations in the Philippines between 1906–1913. Using the later date of 1913 gives a SLAW of 104 years. However, it could be argued that the Philippines lost its *empirical sovereignty* over its territory in 1941 with the strategic retreat of the US and the subsequent conquest by the Japanese Imperial Forces. This is not particularly germane to this discussion of COIN. However, to account as conservatively as possible, this examination includes the latest date possible of 1946 as the start of SLAW.
6. Roger Trinquier, *Modern Warfare: A French View of Counterinsurgency* (Fort Leavenworth, Kansas: US Command and General Staff College, 1961 [1985]).
7. David Galula, *Counterinsurgency Theory and Practice* (Westport, CT: Praeger, 1964).
8. Robert Thompson, *Defeating Communist Insurgency: Experiences in Malaya and Vietnam* (New York: Palgrave Macmillan, 1978).
9. Frank Kitson, *Low Intensity Operations: Subversion, Insurgency, and Peacekeeping* (London: Faber and Faber, 1971).
10. Major General Sir Charles W. Gwynn, *Imperial Policing* (London: Macmillan and Company, 1939).
11. Gian Gentile, *Wrong Turn: America's Deadly Embrace of Counterinsurgency* (New York: Perseus, 2013); Jonathan Monten and Andrew Bennett, "Models of Crisis Decision Making and the 1990–1991 Gulf War," *Security Studies* 29 (2010): 486–520.
12. Andrew F. Krepinevich, *The Army and Vietnam* (Baltimore, MD: Johns Hopkins University Press, 1991); John A. Nagl, *Learning to Eat Soup With A Knife: Counterinsurgency Lessons From Malaya and Vietnam* (Chicago: The University of Chicago Press, 2005); General John Mattis, and Frank Hoffman, "Future War: The Rise of Hybrid Wars," *Pro-

ceedings (November 2005): 132; Heather Gregg, "Beyond Population Engagement: Understanding Counterinsurgency," *Parameters* (2009):18–32; David Killcullen, *The Accidental Guerilla: Fighting Small Wars in the Midst of a Big One* (New York: Oxford University Press, 2009); Michael E. O'Hanlon, *America's History of Counterinsurgency* (Washington, DC: Brookings Institute, 2009).

13. FM 3-24, *Counterinsurgency* (2006) and *Insurgencies and Countering Insurgencies* (2014).
14. Based on my notes regarding assigned lines of operation (LOOs) from COIN operations in Iraq from 2008–2009 and in Afghanistan from 2010–2011.
15. Gil Merom, "Strong Powers in Small Wars: The Unnoticed Foundations of Success," *Small Wars and Insurgencies* 9, no. 2 (1998): 38–63.
16. Martin van Creveld, "On Counterinsurgency," in *Combating Terrorism*, ed. Rohan Gunaratna (Singapore: Marshall Cavendish Intl, 2006).
17. David H. Ucko, "'The People are Revolting': An Anatomy of Authoritarian Counterinsurgency," *Journal of Strategic Studies* 29, no. 1 (2016): 29–61.
18. Daniel Byman, "Death Solves All Problems: The Authoritarian Model of Counterinsurgency," *Journal of Strategic Studies* 39, no. 1 (2016): 62–93.
19. Byman, 62.
20. Ucko, "The People are Revolting"; David H. Ucko and Robert Egnell, *Counterinsurgency in Crisis: Britain and the Challenges of Modern Warfare* (New York: Columbia University Press, 2013).
21. Robert Conquest, *The Harvest of Sorrow: Soviet Collectivization and the Terror-Famine* (Oxford: Oxford University Press, 1986); Victor Dreke, *From el Escambray to the Congo* edited by Mary-Alice Waters (New York: Pathfinder Press, 2002); Jonathan Fenby, *Modern China: The Fall and Rise of a Great Power, 1850 to the Present* (New York: Harper Collins, 2008).
22. Andrew Mack, "Why Big Nations Lose Small Wars," *World Politics* (1975): 175–200.
23. Ivan Arreguín-Toft, "How the Weak Win Wars: A Theory of Asymmetric Conflict," *International Security* 26, no. 1 (2001): 93–128.
24. Merom, "Strong Powers in Small Wars."
25. Jeffrey Record, "Why the Strong Lose," *Parameters* (Winter 2005–2006): 16–31.
26. Andrew James Birtle, *U. S. Army Counterinsurgency and Contingency Operations Doctrine, 1860–1941* (Washington, DC: Center for Military History, United States Army, 1998).

27. Birtle, *U. S. Army Counterinsurgency*; Max Boot, *The Savage Wars of Peace: Small Wars and the Rise of American Power* (New York: Basic Books, 2002).
28. FM 3-24 (2006), 1–13.
29. Steven M. Goode, "A Historical Basis for Force Requirements in Counterinsurgency," *Parameters* (2010): 45–57; James T. Quinlivan, "Force Requirements in Stability Operations," *Parameters* 23 (1995–1996, Winter): 59–69.
30. John J. McGrath, *Boots on the Ground: Troop Densities in Contingency Operations* (Fort Leavenworth, Kansas: Combat Studies Institute, 2006).
31. Goode, "A Historical Basis for Force Requirements in Counterinsurgency"; Michael Lewis, "Replication and Extension: Analysis of "A Historical Basis for Force Requirements in Counterinsurgency," unpublished paper with George Mason University, School for Conflict Analysis and Resolution, 2010; McGrath, *Boots on the Ground*.
32. John S. Brown, "Numerical Considerations in Military Occupations," *Army* 56, no. 4) (2006): 86–87; McGrath, *Boots on the Ground*.
33. Harry G. Summers Jr., *On Strategy: A Critical Analysis of the Vietnam War* (New York: Presidio Press, 1995).
34. Boot, *Savage Wars of Peace*; interview with General Anthony Zinni (USMC, ret) January 16, 2017.
35. Boot, *Savage Wars of Peace*, 310–311; Zinni interview (2017).
36. General David Petraeus (USA, ret), "Counterinsurgency" (lecture, Expeditionary Warfare School, Quantico, VA, April 9, 2015); Rowan Scarborough, "US Troop Withdrawal Let Islamic State Enter Iraq, Military Leaders Say," (July 26, 2015), http://www.washingtontimes.com/news/2015/jul/26/us-troop-withdrawal-let-islamic-state-enter-iraq-m/.
37. *The Small Wars Manual* MCRP 12-15. (1940). (Washington, DC: United States Marine Corps, 1940), 1-1.
38. Richard Caplan, "From Collapsing States to Neo-Trusteeship: The Limits to Solving the Problem of 'Precarious Statehood' in the 21st Century," *Third World Quarterly* 28, no. 2 (2007): 242.
39. For discussion regarding the methods and efficacy of trusteeship forms of relationship, see Boot, *Savage Wars of Peace*; Caplan, "From Collapsing States to Neo-Trusteeship"; Cindy Daase, "International Governance via Shared Sovereignty Arrangements between International (Donor) Organizations and Weak States," *Social Science Research Network (SSRN.com)*, 2011, https://papers.ssrn.com/sol3/papers.cfm?abstract_id=1933224; James D. Fearon, and David D. Laitin, "Neotrusteeship and the

Problem of Weak States," *International Security* 28, no. 4 (Spring 2004): 5–43; Stephen D. Krasner, "Sharing Sovereignty: New Institutions for Collapsed and Failing States," *International Security, 29*(2) (Fall 2004) 85–120; Stephen D. Krasner, "The Case For Shared Sovereignty," *Journal of Democracy* 16, no. 5 (January 2005): 69–83.

40. For discussions on civilian trusteeships and military government, see *The Small Wars Manual* MCRP 12-15 (1940); Rebecca Patterson, *Revisiting a School of Military Government: How Reanimating a World War II-Era Institution Could Professionalize Military Nation Building* (Washington, DC: Ewing Marion Kauffman Foundation, 2011); *Advising: Multi-service Techniques, Tactics, and Procedures for Advising Foreign Forces, MCRP 3-33.8* (Washington, DC: United States Marine Corps, 2009).
41. Patterson, *Revisiting a School*, 5.
42. Boot, *Savage Wars of Peace*; Gravatt, *The Marines and the Guardia*; Alan L. McPherson, A. (Ed.), *Encyclopedia of US Military Interventions in Latin America* (Santa Barbara, CA: ABC-CLIO, 2013).
43. Daase, "International Governance via Shared Sovereignty."
44. Bernard B. Fall, *Street Without Joy: The French Debacle in Indochina* (Mechanicsburg, PA: Stackpole, 2005).
45. See, in particular, chapters 12 and 13 of the *The Small Wars Manual* (1940) for an excellent explanation for the legal requirements to employ US officers in the civilian and military institutions of another state; Boot, *Savage Wars of Peace*; Richard L. Millett, *Searching for Stability: The US Development of Constabulary Forces in Latin America and the Philippines* (Fort Leavenworth, Kansas: US Army Combined Arms Center-Combat Studies Institute Press, 2010).
46. *Measuring Stability and Security in Iraq-Report to Congress* (Arlington, Virginia: US Department of Defense, 2006).
47. FM 3-24 (MCWP 3-33.5) Counterinsurgency (Washington, DC: Headquarters Department of the Army, 2006).
48. Gravatt, *The Marines and the Guardia*, 273.
49. Paul D. Miller, *Armed State Building: Confronting State Failure 1898–2012* (New York: Cornell University Press, 2013), 4–7.
50. Karnow, *In Our Image*, 20–21; David J. Silbey, *A War of Frontier and Empire: The Philippine-American War, 1899–1902* (New York: Hill and Wang, 2007), 74–75.
51. Miller, *Armed State Building*, 14.
52. Hossein Mahdavy, "The Pattern and Problems of Economic Development in Rentier States: The Case of Iran," in *Studies in the Economic His-*

tory of the Middle East, edited by M. A. Cook (Oxford: Oxford University Press, 1970), 428–467; and Hazem Beblawi, "The Rentier State in the Arab World," *Arab Studies Quarterly* 9(4) (1987), 383–398.
53. Beblawi, "Rentier State in the Arab World"; Hazem Beblawi, "The Rentier State in the Arab World," in *The Arab State*, ed. Giacomo Luciani (London: Routledge, 1990); Mahdavy, "Pattern and Problems of Economic Development in Rentier States," 428.
54. Beblawi, "Rentier State in the Arab World," 384.
55. Mahdavy, "Pattern and Problems of Economic Development in Rentier States," 437.
56. Barnett R. Rubin, "Political Elites in Afghanistan: Rentier State Building, Rentier State Wrecking," *International Journal of Middle East Studies* 24, no. 1 (February 1992): 77–99; Rolf Schwarz, "The Political Economy of State-Formation in the Arab Middle East: Rentier States, Economic Reform, and Democratization," *Review of International Political Economy* 15, no. 4 (October 2008): 599–621.
57. Beblawi, "Rentier State in the Arab World," 385–386.
58. Krepinevich, *The Army and Vietnam*, 3.
59. Nagl, *Learning to Eat Soup With A Knife*; Theo Farrell, "Improving in War: Military Adaptation and the British in Helmand Province, Afghanistan, 2006–2009," *Journal of Strategic Studies* 33, no. 4 (2010): 567–594.
60. Carl von Clausewitz, Michael Howard, Peter Paret, and Bernard Brodie, *On War* (Princeton, NJ: Princeton University Press, 1984).
61. John R. Boyd, "The Essence of Winning and Losing," (June 28, 1995), https://web.archive.org/web/20110401221705/http://www.danford.net/boyd/index.htm.
62. John R. Boyd, *Destruction and Creation* (Fort Leavenworth, KS.: US Army Command and General Staff College, 1987).
63. Joel S. Migdal, *Strong States and Weak Societies: State-Society Relations and State Capabilities in the Third World* (Princeton, NJ: Princeton University Press, 1988), 32–33.
64. David Collier, "Understanding Process Tracing," *Political Science and Politics* 44, no. 4 (2011): 823; Stephen Van Evera, *Guide to Methods for Students of Political Science* (Ithaca: Cornell University Press, 1997), 54.
65. Andrew Bennett, and Colin Elman, "Case Study Methods in the International Relations Subfield," *Comparative Political Studies* 402 (February 2007): 170–195; Andrew Bennett, "Process Tracing and Causal Inference," in *Rethinking Social Inquiry*, edited by H. Brady, and D. Collier (Lanham, Maryland: Rowman and Litlefield, 2010); John Gerring, "What

Is a Case Study and What Is It Good for?" *American Political Science Review* 9, no. 2 (May 2004): 341–354; Van Evera, *Guide to Methods for Students*.

66. Dan Slater, and Daniel Ziblatt, "The Enduring Indispensability of the Controlled Comparison," *Comparative Political Studies* 46, no. 10 (2013): 1301.

67. Barbara S. Torreon, *Instances of Use of United States Armed Forces Abroad, 1798–2015* (Washington, DC: Congressional Research Service, 2015).

Chapter 2

The Stumbling State

The Philippines

> The vision for the American colony in the Philippines did have some difference from those of the Europeans and the Japanese. It was conceptualized as a grand and progressive public works project. The Americans' plans were always placed in a 'tutelage' framework rather than the 'peace, order, and justice' mantras of the British and French imperialists. The tutorial was publicized as an attempt to prepare the Filipinos for eventual self-rule, not as a means to perform maintenance on an empire.
>
> –David Ekbladh, *The Great American Mission*[1]

By lunchtime on May 1, 1898, after a little over four hours of battle, the United States Asiatic Squadron under the command of Commodore George Dewey lay at harbor in Manila Bay, the Philippines. Six days after declaring war on April 25, 1898, the US had sailed its Pacific Squadron to Manila Bay and destroyed the entire Spanish fleet in the Far East. Surrounding the American naval squadron was the wreckage of the Spanish Pacific Squadron and the remains of 300 hundred years of Spanish Empire in the Far East. On the close of May 1, 1898, the Spanish lowered their ensign on the flagship of the Spanish squadron, and Americans replaced it with their colors.

In 2004 President George W. Bush coined the phrase "catastrophic success" in Iraq to describe a military success the US was entirely unprepared to exploit or secure.[2] In the Philippines, by May 1, 1898, the US found itself subject to such a catastrophic success. Europeans who saw Dewey sail off to battle the Spanish were certain that the US fleet had no chance of success against a legendary naval power.[3] However, just two and a half months later, the US found itself in possession of a foreign capital with the responsibility to govern and recreate the state there.

The "harsh and philanthropic war" described by John Bass would result in what Rudyard Kipling described as a "savage war of peace" in his ode to the Philippine War.[4] This intervention would produce a Philippine state that would endure longer than any other state developed by the US as a result of COIN intervention by US combat formations. It also produced something else entirely unique for its time. Regarding the Jones Bill of August 1916 that first promised Philippine independence, Stanley Karnow relates, "However flawed, it was a landmark gesture. No other western power at that stage had conceded autonomy to a colony, much less promised it independence."[5] Similarly, Sergio Osmena, the future first vice president of the Philippines, congratulated future president Manuel Quezon, "Your sincere and steadfast efforts have saved your country centuries of suffering that other peoples have to go through on their way to final emancipation."[6] This early promise of independence distinguished American trusteeships from more permanent European versions.

Examining the Philippine Case

Surprised by the catastrophic success of defeating the Spanish so completely at sea, the US was completely unprepared to land forces in the Philippines to engage Spanish forces ashore or hold ground. On May 19, 1898, in what seemed a pragmatic solution, the US unwittingly created and equipped the insurgency it would later fight for control of the Philippines. The US had no material ground combat power immediately available to challenge the Spanish forces ashore in the Philippines. In the weeks before May 19, Commodore Dewey's surrogates recruited Emilio Aguinaldo, a Filipino revolutionary leader who had been exiled to Hong Kong after an unsuccessful attempt in 1896–1898 to gain Philippine independence from Spain. After recruiting Aguinaldo, Dewey's officers transported Aguinaldo to the Philippines and provided him with confiscated Spanish weapons. Dewey then sent Aguinaldo ashore to accomplish on land what Dewey could only do at sea at the time.[7]

Once ashore, Aguinaldo began work immediately. On May 24, 1898, Aguinaldo issued a proclamation creating the Philippine Dictatorial Government. On May 28, the Philippine Dictatorial Government had its first major success in combat of the war, defeating and capturing approximately 270 Spanish Naval Infantry. By June, before the first American soldier arrived from Major General Wesley Merritt's expeditionary force, the remaining Spanish garrisons had been trapped inside their outposts. Those in Manila were entirely besieged on all sides by Aguinaldo's General Antonio Luna and his 30,000 troops.[8]

With the catastrophic success of Manila Bay, US civilian and military leaders began promulgating policy in extremis from May to October 1898. In his account, Aguinaldo reported that Dewey had verbally supported Filipino Independence in July 1898 by relating that the US had no need for further land acquisitions.[9] Admiral Dewey, for his part, was operating without any concrete policy guidelines from President McKinley, other than defeating the Spanish—which had been done faster than anyone had anticipated. And McKinley did not provide any substantive official guidance until October 1898, during the negotiations for the Treaty of Paris.

Nevertheless, Dewey, and later General Elwell Otis, were required to defeat the Spanish trapped inside Manila and throughout the archipelago. The US forces that arrived in July 1898 were woefully unprepared to assume the brunt of the fighting with either the Spanish or the Filipino Nationalists. US forces were overwhelmed early on by the climate and were thrown back by Spanish counterattacks.[10] Further, positions and weapons the early US forces lost as a result of tactical retreats were often recovered and begrudgingly returned by Filipino rebels.[11] Still, on August 13, 1898, Spanish and US forces conferred and contrived a sham battle that would allow the Spanish to surrender to the Americans. The Spanish feared the vengeance of the Filipinos they had oppressed for a little over three centuries.[12] On August 17, 1898, US and Spanish forces performed their sham battle, and US troops entered Manila almost entirely without a fight while forcing the Filipino forces to remain outside the city.

On June 12, 1898, Aguinaldo's government declared its independence. While US forces deployed into theater and negotiations between the US and Spain took place, Aguinaldo undertook a coordinated strategy to prepare for war with the Americans. He cultivated political credibility and maneuvered tactically to achieve positions of advantage. Aguinaldo called for a national assembly in September 1898 to draft a Philippine constitution. The first draft, was an egalitarian document from Filipino lawyer and revolutionary, Apolinario Mabini, who was nicknamed "the brains of the revolution." This constitution provided for a national assembly elected by the people and guaranteed universal suffrage, freedom of the press, religion, and speech. However, the national assembly ultimately settled on a far less egalitarian version written by wealthy *illustrados*, lawyers Felipe Calderon and Felipe Buencamino which called for what they termed an "oligarchy of intelligence." This instrument created an ostensibly Filipino version of the old Spanish system and rejected the more populist constitution written by Mabini. The Calderon and Buencamino constitution ultimately made it far easier for the US to exploit the social cleavages present between the wealthy Filipino elite and the poorer peasantry.

The turning point in US policy came earlier in October 1898 when President McKinley, after months of vacillation, finally decided that the US would take the Philippines as part of the negotiations that resulted in the Treaty of Paris with Spain in December 1898. President McKinley and Secretary of War Elihu Root did not believe the Philippines were capable of self-governance in 1898. Root, who was the principal policy agent for what would constitute postwar governance in the Philippines, investigated and ultimately rejected European and Japanese forms of colonization and imperialism. Whereas he did not believe that Filipinos were able to govern themselves at the time, he did believe that they deserved the same rights as American citizens (excepting the rights to bear arms and enjoy trial by jury). He also insisted that US intervention and occupation should be temporary. President McKinley confirmed the temporary nature of the intervention and the need to affirm US sovereignty. Yet, McKinley also encouraged US officers to adapt American rule to local customs and even prejudices.[13]

By December 1898, President McKinley articulated his policy of benevolent assimilation and had this communicated to the Filipino people. By this policy, the US claimed sovereign control of the Philippines ostensibly to work for the good of the Philippine people. This work on behalf of the Filipino people was effected in much the same way the US was already doing in Cuba. The key difference between the two cases was that the Teller Amendment prohibited the US from annexing Cuba.

COIN THEORIES OBSERVED

During the Philippine-American War and the subsequent Moro Rebellion, both population-centric and enemy-centric COIN methodologies were present, weighted at different times, and effective. On February 2, 1899, Nebraskan Private William Grayson's patrol had a chance encounter at night with a Filipino Nationalist patrol. This contact left several Filipinos dead and began open hostilities between the US and Filipino Nationalist forces. After five months of increasing tensions and positioning, and just two days before the US Congress ratified the 1898 Treaty of Paris, US forces in Manila began a general offensive. For the next ten months, Filipino forces could not keep pace with US operations once the Americans had the initiative, despite possessing solid Filipino leadership, significant numerical advantage, and comparable small arms, at least initially.[14]

The US forces in the Philippines evolved through different tactical methodologies throughout the course of the intervention. US forces began the war by prosecuting a very successful conventional war of maneuver against the Filipino Nationalists from January 1899 until the Battle of Tirad Pass in December 1899. After Tirad Pass, Aguinaldo realized that he could not compete with the US forces conventionally and reverted to a version of what Mao would describe as a Phase II guerilla force.[15] US forces under General Elwell Otis nevertheless continued to execute fruitless large-scale offensive operations despite the Filipino Nationalist turn to smaller scale guerilla operations.

From 1899 to 1900, General Otis not only pursued Aguinaldo's forces but also established local governing councils and a national judiciary. In April 1899, Aguinaldo reached out to Otis to discuss terms for a truce. Otis highlighted proofs of US intentions for predominantly Filipino self-rule under US supervision to the Filipino Nationalist delegates. Otis showed the delegates what was already being done to effect this self-rule in local councils, a nascent Philippine judiciary, and Filipino magistrates. This caused an irreparable rift within the Nationalist leadership when they returned to Aguinaldo and led to the defection of key illustrados, like Pedro Paterno.[16]

June 1900 brought with it the Taft Commission as well as a change in tactics, including martial law under General Arthur MacArthur Jr. Taft and MacArthur's combined policies of attraction and chastisement proved a potent hybrid-COIN strategy.[17] Under Taft's civilian oversight, state building efforts expanded exponentially. This included massive vaccination and sewage programs, as well as the construction of ports, roads, hospitals, and schools. In parallel, MacArthur had foreseen the turn to a "filthy guerilla war of surprise and guile." Subsequently, he proclaimed a strategy of amnesty to those who would surrender and severe chastisement to those who did not.[18] Taft's policies of attraction were executed simultaneously with MacArthur's relentless pursuit of insurgents and the mass resettlement of Filipino citizens into centralized camps—or chastisement.

The Taft Commission was given broad powers to collect taxes, provide for universal education, make and enforce laws, and appropriate funds. The application of these powers was heavily influenced by the advice of Felipe Calderon, who advised that Filipino elites could not be brought along by force. They could, however, be attracted. From Calderon's advice, Taft articulated his policy of attraction. As Karnow observes, "A credible native opposition would deflate Aguinaldo's movement. Over the long run, too, the participation of able Filipinos in the colonial administration would spare the United States the stigma of outright imperialism...From start to finish of America's rule, the Filipinos essentially governed themselves under increasingly light US supervision."[19] Nevertheless, in the short-term, the capture of Aguinaldo had the most pronounced impact on the Filipino Nationalist insurgency.

On the morning of March 14, 1901, Brigadier General Fredrick Funston, four other US officers, and eighty Filipino Macabebe Scouts executed one of the most ambitious raids in modern warfare. American forces had intercepted a dispatch from Aguinaldo and felt certain they could ascertain the location of his command post from it. The patrol conducted an amphibious insertion on the east side of Luzon and began a one-hundred-mile trek through mountainous jungles. They employed a ruse pretending that the Macabebe Scouts were actually insurgents who were bringing Funston and the other American officers as captives to Aguinaldo. The risk to the lives of the US officers and the Macabebes if captured cannot be overstated. This shared risk and mutual trust testifies to the loyalty developed between the Americans and the Macabebe Scouts. Despite near starvation the force penetrated Aguinaldo's camp, and the Americans and Macabebes captured Aguinaldo and marched him back through insurgent territory and to friendly lines. The capture of Aguinaldo did not end the war immediately, but it deflated the insurgency decisively. By July 4, 1902, President Theodore Roosevelt was able to legitimately declare an end to the conflict with the Filipino Nationalists only three and a half years after it had begun.

Population-centric COIN methods were evident in the efforts by the US, and later the Philippine government, to isolate the population from the insurgents. The US and Philippine forces isolated the insurgents geographically, ideologically, and/or morally from their bases of support by relocating populace, addressing grievances, and protecting the populace. Both the Filipino Nationalists and later, after World War II, the communist Hukbalahaps relied on local and regional peasant support for logistics, recruiting, and intelligence. Both the Filipino Nationalists and the Hukbalahaps were starved and/or marginalized into irrelevance by population-centric policies.

The US and Philippine leadership co-opted the Philippine populace from 1899 to 1913 and again from 1945 to 1954. They did this by demonstrating greater legitimacy and empirical sovereignty. The US provided better schools, roads, ports, governance, and essential services. In 1899, the US sought to provide economic development while also either moving the population to defendable enclaves or garrisoning population centers. This is observable in the co-option of the hemp trade, protection of personal property rights, opening of trade overseas, and the use of relocation camps.

Enemy-centric or authoritarian COIN is also observable. Under Generals Otis and MacArthur, we see this policy at work against the Filipino Nationalist forces, particularly with the reapplication of General Order 100 from the US Civil War. According to Francis Lieber's *Guerilla Parties*, the order prescribed what treatment might be permitted for different classifications of combatants, non-combatants, and brigands.[20] Whereas General Order 100 allowed for summary execution and mass detainments, it also proscribed this to a broad array of combatant and non-combatant statuses. Operations under General Order 100 focused on decapitation strikes, clearing operations, raids, ambushes, and surgical strikes. Enemy-centric/authoritarian COIN operations applied intense pressure on Filipino insurgents and required them and their supporters to pay a very high price for continued rebellion.

The efficacy of US COIN efforts is best attested to by the outcomes. First, the evaluation by ardent foe Manuel Quezon is important. He lamented, "Damn the Americans, why don't they tyrannize us more."[21] COIN is essentially warfare among the populace, and it is particularly harsh on those caught in the crossfire. The Philippines, like every other COIN operation the US participated in thereafter, produced some degree of atrocities from both sides. Yet authors like Leon Wolff[22] and Stuart Creighton Miller[23] tend to overstate the racial intolerances/prejudices of the day and American missteps, while failing to also address insurgent excesses. They also tend to side-step Quezon's positive evaluation of American COIN efforts. These authors ignore the fact that perhaps the most noteworthy thing US soldiers from 1899 to 1913 are remembered for by the Filipino populace is teaching them to read and write.[24] Silbey goes to great length to point out that, in spite of the excesses on both sides, what is remarkable is the enduring amity that evolved after the war between Filipinos and Americans.[25]

TROOP RATIOS

Surprisingly, the counterinsurgent-to-population ratios for the Philippine-American War were the lowest of the four cases examined in this study. Yet, the SLAW produced as a result of US intervention was the highest. The general range of minimal counterinsurgent-to-population ratios range between 1:50 and 1:91.[26] At its best, the US counterinsurgent-to-population ratio never got better than 1:100 and averaged only 1:309.

At its height, the US Philippine War saw 74,000 American troops stationed in the Philippines in December 1900, or one US soldier for every 100 Filipinos.[27] The average troop strength for the duration of the conflict hovered at around 40,000, and just two years after the peak strength, only

15,000 US troops remained.[28] With a Philippine population of 7,409,000[29] in 1902, this provided a force ratio of close to 1:100 US troops to Filipino citizens. However, with the Filipino population spread out over 7,000 islands and 115,000 square miles of land, that ratio provided only 1 US troop per 1.55 square miles. Nevertheless, with a max strength of 74,000 and an average of 40,000, US forces did impressively well against an insurgent force estimated at a max strength of 80,000–100,000.[30]

Brian Linn takes the analysis even further. He points out that these troop strengths do not tell the whole story. These numbers do not consider that the average troop strength in theater was only around 40,000 US soldiers. This would provide an average troop-to-population ratio of 1:185[31] and troop-to-insurgent ratio of 1:2–2.5.[32] Moreover, Linn points out that this average number of 40,000 US troops also fails to account for those who were in transit into and out of theater or in some other non-operational status such as sick call. He argues that this would leave only a best-case estimate of 60% of total force, or 24,000 operational troops, available to either hold terrain or conduct operations in the field.[33] Krepinevich calls this foxhole strength.[34] This is important because this lowers the actual useful percentage of US troop to Filipino citizens from 1:100 max at the peak (not adjusting for combat ineffectives) to possibly as low as 1:309 on average (after adjusting for combat ineffectives). Moreover, this provides a further ratio of US troops to square miles of coverage of 1:2.6 at the peak and 1:4.8 square miles on average. McGrath estimates that there was an average of 61 Filipino citizens per square mile.[35]

The Philippine case seems to considerably outperform the expected troop ratios typically identified as minimums. In 34 of the 77 provinces (44%), there were no US military operations, while other areas were under constant attack.[36] It appears, for instance, that the US Philippine War was every bit as violent as Iraq, given the casualty ratios.[37] However, the force ratio for troop to population density was 1:100 at the peak of the war and a low of 1:309 on average. Considering that the most frequently quoted minimum troop-to-population ratios are between 1:50–1:91, the troop ratios in the Philippines fall well short. McGrath does point out the multiplicative impact of indigenous forces as a replacement or amplifying effect of US troop ratios. However, this is of little use in explaining the US's success in terms of SLAW because it was not until 1903, after the defeat of the Filipino Nationalists, that 40% of the combat forces were made up of Filipinos.[38]

PRESENCE OF SANCTUARIES AND CONTINUED COMBAT SUPPORT

The US never foreclosed completely on insurgent sanctuaries in the Philippines. The US and Philippine government did have success in penetrating insurgent sanctuaries, but they never completely interdicted their use by insurgents. The Philippines archipelago consists of over 7,000 islands. Of these, 1,000 islands are inhabited, representing 115,000 square miles of land. It would be a virtual impossibility to entirely foreclose on insurgent sanctuaries. With a maximum potential of 74,000 US troops and a realistic average combat effective force of only 24,000–40,000 US soldiers to fight and hold ground, securing 1,000 islands is even more improbable.

Whereas the US and Philippine forces were not successful at foreclosing the insurgent sanctuaries in Luzon and the Sulu Sultanate, they were able to penetrate these on occasion. The capture of Aguinaldo indicated the ability of the US and Macabebe Scouts to penetrate Aguinaldo's sanctuaries on Luzon, but not to interdict them permanently. Further, US and Philippine Scouts were able to penetrate the Sulu Sultanate on several occasions, culminating in the Battles of Bud Dajo in 1906 and Bagsak in 1913. Despite these successes at penetrating these sanctuaries, insurgent sanctuaries continued to persist.

CONTINUED SUPPORT AFTER WITHDRAWAL

Continued US support after withdrawal or cessation of US combat operations falls short of explaining the long-term success of the Philippine state after withdrawal of US combat forces from combat operations. After World War II, the US did not provide combat support nor combat troops to help defeat the Hukbalahap insurgents.

In terms of money spent during the 1899–1902 conflict, the US-Philippine War cost the US—numbers have been adjusted to account for inflation and to reflect what they would be at the time of this publication in 2019—approximately $11.7 billion[39] for the three years of conflict, or approximately $3.9 billion per year on average.[40] US direct support to the Philippine government during the Hukbalahap Rebellion cost approximately $19.6 billion[41] in military spending in 1948–1954 and non-military spending in 1951–1954.[42] If these direct conflict costs are totaled, then what amounts to ten years of conflict cost approximately $31.3 billion, or approximately $3.13 billion per year. The costs to support the Philippine government after withdrawal and during the Hukbalahap Rebellion is only $19.6 billion for seven years of conflict, or $2.8 billion per year. To put this into perspective, in just FY2017, the Obama Administration requested $44 billion for just one year in Afghanistan.[43]

GOVERNANCE AND SECURITY FORCE DEVELOPMENT

Degrees of Embeddedness in Governance Development

Linn calls the period from roughly June 1899 to early fall 1899 "The Summer Stalemate," and from a purely military and tactical viewpoint, this is an apt description.[44] However, in trying to comprehend what enables a state to endure after US forces have withdrawn, this "Summer Stalemate" was anything but. It was not only the turning point of the US-Philippine War but also the turning point of the history of the modern Philippine state. First, it represents the advent of Filipino governance institutions, such as the Manila magistrates, Supreme Court, and the local governing councils. With these institutions, the Philippine state took its first steps toward self-governance with Americans in the lead. This period constitutes the practical end of the ability of the Filipino Nationalist forces to demonstrate empirical sovereignty. This period would see the grassroots inauguration of the Philippine military institutions which would endure long after US military forces ceased unilateral combat operations.

From 1898 to 1935, the US used institution-inhabiting strategies to replace the Spanish colonial governing institutions. Initially, US agents leveraged trusteeship relationships at the national level politically and at the strategic and tactical levels militarily. These institutions evolved from the Spanish colonial models. US officers adapted these colonial institutions into uniquely American-Filipino institutions as seen, for instance, in the evolution of the Philippine Supreme Court. The US initially employed American military government from February 1899 until June 1900 and initiated the creation of a significant portion of the host-nation governance institutions during this period. From the start, US military and civilian officers encouraged local governance and sought to allow for self-rule to the greatest degree possible. Later, the US agents transitioned into shared sovereignty arrangements at the local levels of government. This provided local councils with great autonomy albeit constrained by US oversight and veto authority. After the creation of the Philippine legislature and promises of independence as early as 1916, US agents transitioned into shared sovereignty arrangements at every level of government. Consequentially, the Philippine government has imperfectly maintained sovereignty over the archipelago for over 71 years since its independence, or eighteen times longer than the US-supported regime in Iraq.

In October 1898, during the negotiations for the Treaty of Paris, President McKinley and Secretary Root doubted the capacity of Filipino elites to govern effectively and secure the Philippines. Filipino elites, like future Supreme Court Justice Cayetano Arellano and others who had fought against Spanish colonial rule, also argued that the Philippines were not ready at that time.[45] Aguinaldo himself acknowledged the real and present danger presented by the German and French warships that appeared on the scene immediately following the demise of the Spanish Fleet. Aguinaldo had expressed on several occasions his gratitude to US naval forces that secured the right of Filipinos to operate Filipino flagged vessels. He was particularly grateful when US ships prevented harassment and boarding by German, French, and Japanese warships. These US efforts to protect Filipino vessels almost led to naval battles between the US and these global powers.[46] It is evident that in these statements, as well as others made by Aguinaldo and other illustrados, they never envisioned an entirely independent Philippine state. Instead, they envisioned a protectorate relationship with the US where they were insulated from international threat by the US but also possessed ultimate veto power over the US regarding Filipino affairs. In essence, Filipino Nationalists wanted sovereignty without responsibility. They wanted US protection without US authority.

US promises of self-rule from the earliest part of the intervention distinguished the American trusteeship in the Philippines from European and Japanese versions. US promises of independence had a deleterious effect on the strength of the insurgency.[47] President McKinley promised that the Philippine Islands would be largely self-governed, albeit under the tutelage of the US, in his proclamation of benevolent assimilation in late 1898. Moreover, McKinley asserted that US tutelage would persist only until such a time as the Philippine state could demonstrate capable self-governance. An essential element of this tutelage, though, would be acceptance of unchallenged US sovereignty during the tutoring period. In effect, the US would not accept responsibility without agency. Less than four months after McKinley's proclamation of benevolent assimilation, the US would specifically articulate the principles that would lead to eventual Filipino self-rule.

In April 1899, the Schurman Commission provided a list of "regulative principles" designed to move the Philippine state toward self-rule as expeditiously as possible.[48] These principles instructed that US agents should attempt to provide a modicum of political autonomy to Filipino citizens as soon as possible, a fair justice system, capable civil servants, and reliable public works. Alongside these provisions, the US would also expect acceptance of their complete sovereignty in order to bring about this state building regimen.[49] The Schurman Commission report stressed that American agents and Filipinos should make common cause in the provision of religious freedom, maintenance and protection of personal property rights, and the "largest practicable measure of home rule."[50] This report established a shared sovereignty arrangement in which the native populace would exercise day-to-day local self-rule while a US officer would possess veto authority.[51] At the national level, the US would maintain its trusteeship until replaced by a capable indigenous authority.

A testament to the seismic impact of this report is in the reaction of Apolinario Mabini, Aguinaldo's "brains behind the revolution." Mabini considered the report to be an ingenious and duplicitous scheme to dupe unsuspecting and gullible natives.[52] The Schurman Commission seemed too good to be true as it promised a degree of self-governance that was not terribly far removed from what Mabini and the revolution was already fighting to achieve. Mabini described the report as an offer of, "the amplest autonomy and the fullest political liberty." He also thought it was a vehicle to allow the US to oppress them at will later.[53] The Schurman Commission's principles would transcend their tenure and would provide the undergirding assumptions for US governance in the Philippines for the next thirty-six years. Ultimately, the Schurman principles would be appropriated and expounded upon by the Taft Commission which arrived in June of 1900.

Almost immediately after the Spanish surrendered Manila, US forces under Major General Otis began to designate indigenous magistrates. Only thirty-three days after the start of the war, Lawton's General Field

Order # 8 issued at Baliuag established the first local representative government in areas under US control by allowing locals to elect a mayor and council. Otis would continue this practice from this point on and even expand it.[54] By August 8, 1899, General Otis, relying on the work of William A. Kobbe, produced the first practical plan for civil government and published this as General Order # 43. General Order # 43 directed the president, vice president, and members of town and provincial councils to collect light taxes. These taxes were for the purpose of establishing schools and public works projects; these councils were required to provide a monthly accounting to US garrison commanders to prevent corruption.[55]

Revolutionary change also took place among the previously nonexistent Philippine judiciary. In recording its own history, the Philippine Supreme Court relates that "Secretary of State John Hay, on May 12, 1899, proposed a plan for a colonial government of the Philippine Islands which would give Filipinos the largest measure of self-government."[56] Further, "On May 29, 1899, General Elwell Stephen Otis, Military Governor for the Philippines, issued General Order No. 20, reestablishing the *Audiencia Teritorial de Manila* [...] Thus, the reestablished *Audiencia* became the first agency of the new insular government where Filipinos were appointed side by side with Americans."[57] The new Supreme Court consisted of American and Filipino jurists under the leadership of its first Chief Justice, a Filipino jurist Cayetano Arellano. This was particularly astounding in light of the fact that Spanish colonization disdained, and even at times prohibited, the service of indigenous priests.

General Otis, as military governor, was a lawyer by profession and did not solely provide for the reestablishment of a judicial system but for a larger legal system with Filipinos operating within it. General Otis had prescribed the deputization of an indigenous police force beginning in Manila in July 1899. According to a Report of War Department record in 1900, these native police began with 426 constables and grew to 625 by the year's end. This force made 7,442 arrests, including three insurgent generals.[58] With the support of this native police force, insurgents could not move about in Manila without fear of capture by Filipino police.

June 1900 was a watershed moment where the Philippine state transitioned from an American military government to an American civilian trusteeship under US Governor Willian Howard Taft. Several critical events took place during and after this transition that would solidify the existence of the modern Philippine state. First, the Spanish-era Supreme Court transitioned into its modern form in June 1901.[59] The second event, in 1907, was the creation of the Philippine Legislature, the first popularly elected national legislature in the Far East. Third, with the appointment of Governor Francis Burton Harrison in 1913, much of the bureaucracy of the Philippine government began to be dominated by Filipinos, and in 1916, the US Congress began to construct a pledge of eventual independence for the Philippine state. Fourth, in an unusual turn of events, the US Congress voted to grant the Philippines full independence in 1933, which the Philippine legislature ironically rejected. Fifth, in September 1935, President Quezon was elected as the first president of the Philippine commonwealth, and the Philippines gained their independence in 1946. With these transitions, the US had gradually relocated sovereignty into Filipino hands from the grassroots level up.

It is hard to overstate how revolutionary these trends toward self-government were for the times. These trends toward self-government are particularly revolutionary when compared to the European and Japanese alternatives of the day. The Spanish friars had refused Filipinos entry into the priestly orders and had discouraged the promulgation of Spanish as a lingua franca to prevent archipelagic unity for 300 years.[60] From 1887 to 1951, the French did not allow for Vietnamese to be commissioned anything higher than junior officers in its Indochinese colonial forces.[61] As late as 1994, indigenous military forces like the Ghurkas were still commanded by British officers.[62] Additionally, the independence of European and Japanese colonies was most often granted only after armed revolt and protest, and these did not really begin to take shape until long after the Philippines received its independence. Lastly, European colonial relationships were designed to create a permanent dependence because European colonization was designed with no termination in view.[63] In contrast, the American effort in the Philippines was designed to allow US forces to turn sovereignty over to the Filipinos.

Degrees of Embeddedness in Security Development
From 1898 to 1913, US military officers deeply embedded themselves in the Philippine security forces. The forces these US officers developed would maintain a high degree of security for over a century afterward. The most impressive aspect of the US's efforts was the ability to amalgamate Filipinos from backgrounds riven with sectarian strife into a uniquely Filipino organization led by a neutral US officer. The resultant Philippine military has been able to defeat several significant internal threats and is considered one of the most trusted organs of the Philippine state.[64] The Philippine security forces were created from scratch as encadred units comprised Filipino soldiers and constables led by US officers. Whereas there were formal mechanisms of training, such as the Philippine Military Academy and the US Military Academy, the bulk of American martial culture was transferred and leadership was taught through daily interaction and example.

The Macabebe Scouts and later the Philippine Scouts were very well regarded by every level of the US military command in the Philippines. They were not only largely responsible for the ultimate capture of Aguinaldo but also for the destruction of the Filipino Nationalist forces and the crushing of the Moro revolt. Yet, the creation of this force was not by design but a ground-level adaptation led by an unusual US Army lieutenant.

1st Lieutenant Matthew A. Batson was certain it would take 100,000 US soldiers just to hold the island of Luzon, to say nothing of the other 999 inhabited islands. Batson already had a profound respect for the Filipinos upon his positive reception in the province of Pampanga by the Macabebes. The Macabebes, like many indigenous forces who support foreign forces, were an ethnic minority with significant enemies among the majority Tagalogs. The perpetual discord between these two groups led the Macabebes to support the Spanish first and later the Americans. For their service to the Americans, many of the Macabebe villages were destroyed and the villagers raped and murdered by Tagalog nationalists.[65] Lt. Batson proposed the incorporation of the Macabebes into US operations.

Batson initially received little encouraging feedback, but by September 1899, Otis' 8th Corps was running short of combat power. Between pursuing Aguinaldo and having to defend previously cleared areas, the American leadership was open to novel approaches. On September 10, 1899, Batson raised his first company of 108 Macabebe Scouts.[66] These scouts were of such superlative quality that he was ordered to raise another company on September 21 and a third company on October 6.[67] Despite numerous critiques of US forces by American officers in the US archives, there appear to be no negative references to the capacity and bravery of these indigenous forces in the historical record, at least none that I was able to locate. US officers in the Philippines did not always agree on the capacity of many other US troops or units, but they seem to agree unanimously on the capacity of the Macabebe Scouts led by Batson and his fellow American officers.[68] What is more, Major General Lawton noted in an October 18, 1901, telegram to his chief of staff that the Macabebe Scouts had started such a tradition of excellence that they were often seen as better than the average American soldier:

> Macabebe scouts doing excellent service and are proving model soldiers for this service. *They are worth twice their number of our inexperienced men*...At least let me have a full battalion. [69]

Led by American officers, the ad hoc Macabebe Scouts, about the size of a small battalion, destroyed a Filipino Nationalist battalion and defeated another company on October 29, 1899, by themselves. In August 1901, the Scouts would form the core of what would later be designated the Philippine Scouts. However, as the war progressed, a single ethnic Macabebe composition of the Philippine Scouts was no longer desirable and gave way to the incorporation of Tagalogs into the Philippine government's service. Moreover, with Aguinaldo's capture and subsequent acceptance of US sovereignty, many other insurgents surrendered and were added to the rolls of the Scouts. One key example of this is General Juan Cailles who took over for Aguinaldo but was captured on June 1, 1901, and joined the US operations against his former comrades.[70] As such, the Philippine Scouts began to take on the broad array of diverse ethnicities and backgrounds within the archipelago. Despite his initial reticence to recruit indigenous forces, General Arthur MacArthur relented, and by January 1902 the Scouts had more than 5,500 members.[71]

The inclusion of diverse ethnicities in the Philippine Scouts, that began in 1901, is instructive for understanding how a foreign force might recruit, man, train, and equip an indigenous military in the face of historic sectarian strife. The Scouts were originally made up of an ethnic minority, the Macabebes. The Tagalog majority had formerly oppressed the Macabebes, so they had reason to loathe the Tagalogs. The Tagalogs also had good reason to hate the Macabebes for their collaboration first with the Spanish and then with the Americans. The ability to amalgamate these two ethnicities into a disciplined fighting force became even more impressive when Moro Muslims were also added to the rolls. Moro Muslims had been at war on and off with both the Spanish and the majority Catholic Tagalogs before the US intervention. Nevertheless, under the leadership of foreign American officers, this heterogenous force was able to fight a complex COIN campaign against the Moro rebels in the Sulu Sultanate. The Philippine Scouts were instrumental in defeating the Moro Rebellion at Bud Bagsak in 1913.[72] This provides some insight into the potential value of American officers commanding these extremely diverse military units in the face of bloody sectarian strife. The neutral American officers appear to have been able to unite combat formations despite historic ethnic and religious sectarian strife in ways that indigenous officers might have been unable to do.

On August 18, 1901, as a result of Act #175 of the 2nd Philippine Commission, the Philippine Constabulary became operational. Similar to French gendarmes, they combined the powers of police officer and soldier depending on the threat.[73] This is remarkable because the US officers in the Philippines had no prior experience with a gendarmerie or constabulary but only with discrete military and police forces. What is also remarkable is that from the start, the Philippine Constabulary allowed Filipinos to serve as commissioned officers and immediately commissioned two such officers upon the inauguration of the organization.[74] By way of comparison, the British did not begin to localize or transition the officer corps of their colonial forces like the Royal Malaysian Special Branch until the Malaysian Emergency in 1948–1960, over a century after they began to colonize Malaysia.[75] By the end of 1901, or only half a year later, the Philippine Constabulary had grown to 180 constabularies and was one of the most effective forces in separating insurgents from their base of support.[76]

By 1906, US units and a small detachment of Philippine Constabulary defeated the Moros during the Bud Dajo campaign. In early 1899, Filipinos made up none of the forces defending the Philippines. By 1903, Filipinos from a variety of ethnic and religious backgrounds would make up 40% of the US combat forces in the Philippines. Whereas the US still provided the bulk of the combat forces in 1906, the percentage of Filipino troops was nearing parity.

By the end of the Moro Rebellion with the Battle of Bud Bagsak in 1913, a force of 1,200 Moro Scouts, Philippine Scouts, and Philippine Constabularies had defeated a Moro force of around 500.[77] General Pershing later remarked to his wife that this was some of the fiercest fighting he had ever witnessed despite his long combat record, which included grisly combat in the trenches of World War I. Nevertheless, led by the American officers, the combined force acquitted itself well. The result of this action was to bring the Sultanate of Sulu under American control. From 1913 on, it would be an entirely Philippine military and constabulary operation to maintain the peace without US combat formations or combat support. Whereas the US forces stationed in the Philippines likely dissuaded regional and global hegemons from invading, US combat support and combat-service support were not required by the Philippine government to quell internal insurgencies ever again. The enduring success of these Filipino forces is impressive, given the sectarian violence that separated them in daily life before the Spanish American War.

Another method the US military employed to develop the Philippine military and transfer martial culture was the creation of the Philippine Military Academy. The Philippine Military Academy operated largely along the same lines as the US Military Academy. In 1905, an officer's academy was established as a school for the Philippine Constabulary, and over the years it routinely moved locations until it became the Philippine Military Academy in 1936. From 1905 to 1936, what would become the Philippine Military Academy was led by sixty-one superintendents. Of these, the first thirty superintendents were all American officers. This produced a high and enduring level of integration between the US and the Philippines, even to the degree that it is still enshrined in US law.[78] Thus, the Philippine Military Academy and US Military Academy both provided further opportunity for enduring symbiotic adaptation even as US officers transitioned out of embedded roles within the Philippine security forces.

The outcome of US efforts to develop Philippine governance and security institutions through institution-inhabiting strategies is impressive. Without the Macabebe Scouts led by US officers, the capture of Aguinaldo would have been impossible. Additionally, US efforts to create a system of self-rule from the start of its intervention were entirely novel for their time. As a result, the Philippine security forces began to police Manila within months of the start of the war and even captured three insurgent generals. Within only four years, the Philippine Scouts and Constabulary had replaced 40% of US forces from the bottom up. By

1906, combined Filipino and American forces were able to penetrate the Sulu Archipelago and defeat the Moro insurgents. Within thirteen years, the Philippine security forces were almost entirely comprised of Filipino citizens though officered by Americans. Moreover, the government and security institutions created by the American intervention have been able to survive since 1946 in spite of challenges by further communist and Muslim insurgencies.

Transition to Self-Rule First Test-The Hukbalahap Rebellion 1948–1954

The Hukbalahap bong Mapagalayang Bayan (translates to People's Liberation Army), Hukbalahap, or Hukbalahap Rebellion began in August 1948. The Hukbalahap bong Mapagalayang Bayan derived its origins from the early formation of the Filipino communist party in 1930 by American communists who infiltrated the US mission to the Philippines in the early twentieth century. However, the party was forced underground after the Philippine legislature declared it illegal in 1932. In 1938, the Filipino socialist and communist parties unified to later become the Hukbalahap bong Mapagalayang Bayan. In a fashion similar to the illustrados of the Filipino Nationalist movement, the Hukbalahap intelligentsia was disproportionately made up of wealthy and educated elites. However, its most important military leader, Luis Taruc, was a genuine peasant leader.

After World War II, the Hukbalahaps were seen as dangerous communist insurgents by US forces. American intelligence relied on the assessment of US guerilla leaders and other Filipino elites. These assessments and the growing control of communism in China and North Korean were concerning.[79] From 1946 to 1948, Philippine government forces, Hukbalahaps, and vigilantes took turns conducting violent strikes against one another. Despite being rejected by General MacArthur during World War II, the Hukbalahaps had operated very effectively in support of US efforts to liberate Luzon in 1945. They were not, however, accorded the same trust as other Filipino guerila forces because of their ties to communism. Still, the group participated in the 1946 elections and won six seats in the

Philippine legislature.[80] This election, however, did not end the threat of violence, and President Roxas officially declared the Hukbalahap movement illegal and subversive just before his death in 1948.

From August 1948 to early 1951, the Hukbalahap Rebellion grew to between 11,000 to 17,000 active insurgents, 50,000 reservists, and potentially 2,000,000 peasant supporters depending on different estimates.[81] Hukbalahap forces possessed an almost entirely regional cellular organization in much the same way Aguinaldo's forces had half a century prior. This helps explain the difficulty the Philippine state had in decapitating Hukbalahap leadership. It also explains the Hukbalahap's inability to become a significant national force.

The Hukbalahap Rebellion had its most significant operational year in 1950. President Roxas' "Mailed Fist," enemy-centric COIN strategy in 1948 increased popular bitterness toward the Philippine government and support for the Hukbalahaps. The Philippine security forces' use of indiscriminate violence quickly alienated much of the rural populace. In contrast to Roxas' excesses, the next Philippine President, Elpidio Quirino's overly accommodating policies did little to convincingly attract insurgents with promised reforms or to chastise them through tactical operations. As this battle for an effective strategy continued, the US began to leverage its diplomatic pressure to compel the Philippine government to change course.

As early as 1949, and certainly by the time of the publication of a report commissioned by the US Joint Chiefs of Staff in May 1950, the US saw the collapse of the nascent Philippine government as potentially imminent.[82] Despite investing close to $20.84 billion from 1945 to 1950 (in 2019 dollars),[83] the Truman Administration threatened to back away from the Philippine government unless it undertook radical reforms.[84] Whereas President Quirino was loathe to pursue genuine reforms, even in the face of an existential internal threat, he realized that US aid depended on his acquiescence. Acquiescence to this coercive influence gave rise to Ramon Magsaysay and Edward Lansdale as key players in the Hukbalahap Rebellion.

In a replay of attraction and chastisement, which was itself a replay of moderation and reprisal in the US Civil War, Landsdale, a CIA officer, acted as a strategic advisor to Philippine Secretary of National Defense Ramon Magsaysay. On Landsdale's advice, Magsaysay eschewed large-scale conventional operations that disaffected the populace and had little to show for defeating the insurgents. Instead, Magsaysay leveraged surgical special forces operations like those of Force X and the C Company of the 7th Brigade Combat Team to target key insurgent leaders for capture/kill missions.[85] He also made it a point to tour Philippine military and police positions constantly and demonstrate leadership from the front. The increased supervision allowed Magsaysay to get rid of incompetent and predatory officers. It also increased morale by demonstrating wise use of US military aid to modernize the Philippine forces. The improvement of army morale and pay moved the Philippine Army away from many of its previous excesses and provided a stark contrast to the continued Hukbalahap excesses.

Magsaysay coupled targeted military operations and improvement in Army capacity (chastisement) with genuine attempts at reform (attraction). Whereas some of these reforms were derided as mere propaganda, they were nevertheless received and believed by the populace. Hukbalahaps who surrendered did in fact receive plots of land. However, in order to take ownership of the land, they had to actively work against their former Hukbalahap comrades. Only a small percentage of those Hukbalahaps who surrendered actually received land. Magsaysay's land reform courts, clinics, and other development projects won a great deal of goodwill to the cause of the Philippine government, at least until Magsaysay's death in a plane crash in 1957. This goodwill was evident in the 1951 elections when, despite intense Hukbalahap pressure encouraging a general peasant boycott, a record 4,000,000 voters nevertheless participated.[86]

The combination of attraction and chastisement by the Philippine government, under the direct leadership of Magsaysay, had a decisive effect on insurgent morale. Many former insurgents surrendered and turned against their former comrades in much the same way the Filipino Nationalist forces had done when fighting the Americans. Moreover, just as the internecine fighting among the Filipino Nationalists had severely crippled their efforts, differences between the communist elites and Luis Taruc weakened the Hukbalahap. Fearing the communist elites might liquidate him, Taruc began overtures to Magsaysay. Upon his election to the Presidency in 1953, Magsaysay pounded the final nail in the Hukbalahap coffin. Magsaysay encouraged the populace to share their grievances with the government, which delegitimized the Hukbalahap's role as arbiter between the government and the people. This, and rumors of communist plots against his life, compelled Taruc to surrender to Philippine forces in May 1954. Taruc submitted to the Philippine government and accepted three lifetime sentences, of which he only served fourteen years.[87]

Without the direct intervention of US combat forces, the provision of combat support, or substantial combat-service support, the Philippine government successfully defeated an internal insurgency. Whereas the inaugural test of Philippine independence was a convoluted one, it was no more so than the US's own COIN efforts in the US-Philippine War. In fact, the Hukbalahap Rebellion might well be considered a mere replay of the American and Filipino Nationalist story arcs from half a century prior. Ramon Magsaysay and the Philippine government had defeated a guerilla movement that had been in operation for six years longer than the newly independent Philippine government. The Philippine government succeeded with only a handful of US advisors, minimal economic support, and no US combat support. From 1933 on, the Philippine government demonstrated a capacity for governance and empirical sovereignty unique among the states the US has intervened in with combat formations during COIN. Philippine security forces developed by American officers have been able to provide a high degree of stability for over 100 years.[88]

Conclusion

From 1898 to 1913, the US deployed trustee forms of institution-inhabiting strategies, such as military governance, American civilian governorships, and encadrement, to develop Philippine security and governance institutions. These institution-inhabiting strategies reflected relationships of high degrees of embeddedness. This produced host-nation security forces that were able to act as the senior partner in an oligopoly of violence for over a century. The Philippine state has also been able to weather an incredible number of natural and man-made disasters and crises. This has produced what I describe as a stumbling state, characterized by low democracy and high stability. Ultimately, American civilian and military officers were replaced in a bottom-up, grassroots fashion. This ensured the continuity of the martial culture that was symbiotically produced from the combination of American culture and the Filipino compadrazgo system.

The US COIN strategy of attraction and chastisement was a potent combination. It reflected what contemporary COIN theory would describe as a balanced or hybrid form of COIN methods. Attraction and chastisement relied upon dynamic measures of population-centric and enemy-centric COIN. This methodology decisively defeated the Filipino Nationalist insurgency in 1902 as well as two more Muslim insurgencies in 1906 and 1913. However, what is impressive about the success of the Philippine state against the Hukbalahap Rebellion of 1948–1955 was that this was the first instance where the nascent Philippine state had to win without US leadership or combat support. However, the effective application of COIN tactics does not explain how the US created a host-nation government capable of defeating this Hukbalahaps.

Similarly, the other tactical concerns of troop ratios, interdiction of adversary sanctuaries, and continued combat and economic support also fall short in explaining Philippine state longevity after withdrawal. US troops either had parity with the insurgents or, more often, were significantly outnumbered. The technological gap between these two forces did not provide the US with much of an advantage. The rifles and machine guns were not decisively different. The US did possess naval gunfire assets. However, this was only valuable along the coast. The US also had limited artillery. However, given the difficulty of transporting artillery pieces across the unhospitable terrain of Luzon and the dispersed tactics of the Filipino insurgents later in the war, US artillery does not appear to have been a decisive element in the US success. Furthermore, the ratio of US troops to population was substantially below the levels predicted to be required for successful COIN. And yet, the Philippine state persisted. Moreover, insurgent sanctuaries, though penetrated, were never fully interdicted. And still, the Philippine state persisted. Lastly, during the Hukbalahap Rebellion, US support was reduced to a handful of advisors and only a fraction of the cost of the US's wars in Iraq and Afghanistan.

In contrast, governance and security force development strategies utilizing high degrees of embeddedness created Philippine institutions able to persist long after the withdrawal of US units from combat roles. The leveraging of institution-influencing strategies in developing governance institutions resulted in a more competitive government than the insurgents could provide. It also either modified or entirely replaced colonial institutions in local governance, the rule of law, and the legislative, judicial and executive functions of government. The use of trusteeship forms of relationships initially produced a system of government that, while not perfect, could outperform the system the Filipino Nationalists could provide at the same time. When the illustrados of the Filipino Nationalists rejected agrarian reform and the more egalitarian Mabini Constitution in favor of the elitist Calderon Constitution, this provided a significant opportunity for the American led government. With the promise of eventual national self-rule and a demonstrated preference for Filipino village and provincial self-rule, the American system offered Filipinos better, if not perfect, opportunities for democratic representation and participation.

Whereas some scholars have pointed out that, despite years spent developing Philippine organs of governance and security, the US failed to produce an exact clone of American representative democracy.[89] This stems from a fundamental misunderstanding of the goals of the US and the specious assumption that this would ever be possible or even valuable. President McKinley's stated goal was to provide the Philippines with the best government possible, to bring them peace and prosperity, to "do our best to help our little brown brothers" and "to take the archipelago and 'to educate the Filipinos, and uplift and Christianize them, and by God's grace do the very best we could by them...'"[90] Whereas McKinley's language is decidedly patronizing, jingoistic, and anachronistic, he was also uniquely farsighted for his time. McKinley ordered his agents to conform American political and military cultures to the "local customs and even prejudices."[91] Therefore, from the presidency down, the goal was not to create an American clone, but rather a Filipino version of American political and military cultures.

American officers embedded within the Philippine organs of government and security forces were compelled to adapt symbiotically the American political, economic, and martial culture within the compadrazgo system. Fulfilling McKinley's mandate to adapt US democracy and martial culture required a symbiosis of US officers and Filipino soldiers and bureaucrats within Filipino institutions. Karnow and Stuart Miller seem to argue that anything short of an American democratic and capitalist clone should be considered an American failure to export itself.[92] Yet, McKinley's stated goal from the start was a version of US democracy, capitalism, and martial culture that was discretely adapted to "local customs and even prejudices."[93]

Politically, the American-compadrazgo system in the Philippines looks, at its worst, like the Tammany Hall system in New York. The compadrazgo system was built on centuries of clan and family lines through marriage and other relations, rather than strictly political, ethnic, or ideological connections. American officers were able to embed within this system and amalgamate political and military cultures. Resultantly, the persistence of the Philippine state after the US ceased combat operations is impressive. Unlike instances of COIN intervention after 1950, the US effort in the Philippines produced a capable albeit stumbling state and host-nation government that has persisted for over seventy-one years since independence.

Sectarian strife is ubiquitous in internal war, and the ability for a foreign military to integrate diverse and acrimonious groups into one cohesive security force is remarkable. Moreover, the ability to create host-nation security forces that were able to provide security for over a century is also noteworthy. This is particularly so in light of the inability of US trained host-nation security forces after 1950 to do this for longer than four years.

US forces developed these host-nation security forces through a trusteeship form of relationship called encadrement. The combination of US leadership with local cultural knowledge produced an efficacious combination. What was extraordinary was the merging of formerly bitter sectarian enemies, the Tagologs and the Macabebes, into the same security forces. Even more impressive was the addition of Muslims from the Moro lands to the same fighting force and their ability, under US officership, to decisively defeat the Moros at Bud Bagsak in 1913.

With respect to the impact of the compadrazgo system on the military, its effects can be observed in several important instances. Despite being the son of the American general who had defeated the Filipino Nationalist movement, General Douglas MacArthur is genuinely revered by the Filipino people. And, his promise to return after the loss of Bataan was a powerful element influencing Filipino resistance. According to Karnow, MacArthur was seen as the patrón of the Filipino people.[94] This patronage system also presented itself again in the relationship between Colonel Edward Geary Landsdale and Ramon Magsaysay during the Hukbalahap Rebellion. The result of the symbiotic adaptation between the Americans and Filipinos resulted in the creation of a constabulary organization which the US forces had no previous experience with.

Outcomes and Regime Longevity

In analyzing the US's efforts to develop the government of the Philippines and its security forces, four outcomes are visible. Firstly, the US succeeded in developing a high degree of security of significant duration. Secondly, the US succeeded in creating an effective government, albeit beset by the cultural constraints, of the compadrazgo patronage system. Thirdly, the nascent US-Filipino government produced a semi-effective, low democracy. Fourthly, the US produced what this study describes as a stumbling state. The concept of a stumbling state is consequential because it elucidates an idea of an imperfect democracy beset by natural and man-made disasters but persisting and governing nonetheless. Even as it would be difficult to argue that the Philippines has been successful in "getting to Denmark,"[95] it has nevertheless produced dramatically better results than states the US has intervened in since 1950. Given this, the Philippine case warrants analysis for what lessons it might offer for interventions after 2018.

Notes

1. David Ekbladh, *The Great American Mission*, 20 (Princeton, NJ: Princeton University Press, 2010).
2. "Bush Calls Iraq Invasion a 'Catastrophic Success'," (August 30, 2004), http://www.foxnews.com/story/2004/08/30/bush-calls-iraq-invasion-catastrophic-success.html.
3. Karnow, *In Our Image*, 102–103.
4. Ibid., 155; Rudyard Kipling, "White Man's Burden: The United States and the Philippine Islands," *McClures Magazine* (1899).
5. Karnow, *In Our Image*, 274.
6. Ibid.
7. Emilio Aguinaldo, *True Version of the Philippine Revolution* (1899) (New York: Valde, 2009), 11; Brian Linn, *The Philippine War, 1899–1902* (Lawrence, KS: University of Kansas Press, 2000), 20–21; Silbey, *War of Frontier and Empire*, 41–42.
8. Aguinaldo, *True Version of the Philippine Revolution*, 15.
9. Ibid.
10. Aguinaldo, *True Version of the Philippine Revolution*; Karnow, *In Our Image*; Linn, *The Philippine War*; Silbey, *War of Frontier and Empire*.
11. Aguinaldo, *True Version of the Philippine Revolution*, 25.
12. Karnow, *In Our Image*; Linn, *The Philippine War*; Silbey, *War of Frontier and Empire*.
13. Karnow, *In Our Image*, 170.
14. There is some debate regarding differences between the quality of rifles used by both sides and access to ammunition. However, these were not decisive. What appears to have been more decisive was the operational culture of the American forces in conventional war. See Karnow, *In Our Image*; Linn, *The Philippine War*; and Silbey, *War of Frontier and Empire*.
15. "Following Phase I (organization, consolidation, and preservation) and Phase II (progressive expansion) comes Phase III: decision, or destruction of the enemy. It is during this period that a significant percentage of the active guerrilla force completes its transformation into an orthodox establishment capable of engaging the enemy in conventional battle." Quoted from Mao Tse-tung, *On Guerilla Warfare*, FMRP 12-18 (Washington, DC: Headquarters United States Marine Corps, 1961), 21.
16. Karnow, *In Our Image*, 158.

17. Ekbladh, *The Great American Mission*; Boot, *Savage Wars of Peace*; Karnow, *In Our Image*; Linn, *The Philippine War*; Silbey, *War of Frontier and Empire*.
18. Karnow, *In Our Image*, 157.
19. Ibid., 174.
20. Francis Lieber, *Guerrilla Parties: Considered with Reference to the Laws and Usages of War* (New York: D. Van Nostrand, 1862).
21. Boot, *Savage Wars of Peace*, 125.
22. Leon Wolff, *Little Brown Brother: How the United States Purchased and Pacified the Philippine Islands and the Century's Turn* (Garden City, New York: Doubleday, 1961).
23. Stuart Creighton Miller, *Benevolent Assimilation: The American Conquest of the Philippines, 1899–1903* (New Haven, CT: Yale University Press, 1982).
24. Boot, *Savage Wars of Peace*, 125–128; Karnow, *In Our Image*, 201–208; Linn, *The Philippine War*, 327; Silbey, *War of Frontier and Empire*, 210–212.
25. Silbey, *War of Frontier and Empire*, 210.
26. Brown, "Numerical Considerations in Military Occupations"; FM 3-24 (2006) 1–13; Goode, "A Historical Basis for Force Requirements in Counterinsurgency"; Lewis, "Replication and Extension"; McGrath, *Boots on the Ground*, 109; Quinlivan, "Force Requirements in Stability Operations."
27. Linn, *The Philippine War*, 325.
28. Timothy K. Deady, "Lessons from a Successful Counterinsurgency: The Philippines, 1899-1902," *Parameters* (Spring 2005): 66.
29. The Philippines: Historical Geographical Data of the Whole Country, http://www.populstat.info/Asia/philippc.htm.
30. Linn, *The Philippine War*, 325.
31. 40,000:7,409,000 US troops to Filipino populace.
32. 40,000:80,000-100,000 US troops to insurgents.
33. Linn, *The Philippine War*, 325. Insurgents would most likely have some undetermined portion of their force in a non-operational status as well.
34. Krepinevich, *The Army and Vietnam*, 236.
35. McGrath, *Boots on the Ground*, 8.
36. Linn, *The Philippine War*, 185.
37. This does not factor in Wounded in Action (WIA) as Post-traumatic Stress Disorder (PTSD) and Traumatic Brain Injury (TBI was not a measurable statistic in 1902 which would significantly adjust the numbers).

38. Edward M. Coffman, "Batson of the Philippine Scouts," *Parameters* (1977): 72.
39. $400 million in 1902, www.in2013dollars.com.
40. Allan Reed Millett and Peter Maslowski, *For the Common Defense: A Military History of the United States from 1607–2012* (New York: Free Press, 1994), 316–317; Linn, *The Philippine War*, 219.
41. $2.091 billion in 1954, www.in2013dollars.com.
42. Christian Gerlach, *Extremely Violent Societies: Mass Violence in the Twentieth Century* (Cambridge: Cambridge University Press, 2010), 216.
43. Neta C. Crawford, *US Budgetary Costs of Wars Through 2016* (Providence, Rhode, Island: Watson Institute-Brown University, 2016), 3.
44. Linn, *The Philippine War*, 117.
45. Karnow, *In Our Image*, 153.
46. Aguinaldo, *True Version of the Philippine Revolution*, 17 and 19; Karnow, *In Our Image*, 125.
47. Linn, *The Philippine War*, 109.
48. Dean Conant Worcester, George Dewey, Elwell Stephen Otis, Charles Denby, Jacob Gould Schurman, *Report of the Philippine Commission to the President January 31, 1900* (Washington, DC: United States Philippine Commission 1899–1900, 1901), 5.
49. Karnow, *In Our Image*, 151.
50. Ibid., 152.
51. Daase, "International Governance via Shared Sovereignty."
52. Ibid.
53. Ibid.
54. Teodoro Agoncillo, *Malolos: The Crisis of the Republic* (Manilla: University of the Philippines Press, 1997), 390; Karnow, *In Our Image*; Linn, *The Philippine War*, 114.
55. Linn, *The Philippine War*, 130.
56. A Constitutional History of The Supreme Court of The Philippines, (n.d.), http://sc.judiciary.gov.ph/aboutsc/history/index.php.
57. Ibid.
58. Linn, *The Philippine War*, 128.
59. A Constitutional History of The Supreme Court of The Philippines.
60. Karnow, *In Our Image*, 64–65, 208; Silbey, *War of Frontier and Empire*, 210–212.
61. James Lawton Collins, *The Development and Training of the South Vietnamese Army 1950–1972* (Washington, DC: Department of the Army, 1974), 15.

62. Based on my personal training with the Ghurkas in 1994 in Hong Kong and briefs provided by Ghurka regimental officers. For another discussion concerning the transition of Ghurka leaders to commissioned officer status, see J.P Cross and Buddhiman Gurung's *Gurkhas at War: Eyewitness Accounts from World War II to Iraq* (Barnsley, England: Greenhill Books, 2002).
63. Fearon and Laitin, "Neotrusteeship and the Problem of Weak States," 7; Miller, *Armed State Building*, 4–7.
64. Even during the challenges like the communist uprisings of the 1980s and Marcos authoritarianism, the military maintained a high degree of confidence, see Marites Danguilan-Vitug, "Filipino military rides popularity wave. But key task is to maintain people's support, officers say," (March 18, 1986), https://www.csmonitor.com/1986/0318/opma.html.
65. Coffman, "Batson of the Philippine Scouts," 68–72.
66. Ibid., 71; Linn, *The Philippine War*, 128.
67. Ibid.
68. Boot, *Savage Wars of Peace*; Coffman, "Batson of the Philippine Scouts"; Linn, *The Philippine War*; Silbey, *War of Frontier and Empire*.
69. *Telegram From General Lawton to Chief of Staff Schwan*, October 18, 1901 (National Archives Washington, DC: RG395, Box 2, File 5), 47 (italics added); Coffman, "Batson of the Philippine Scouts," 68–72; Linn, *The Philippine War*, 128 and 143–144.
70. Karnow, *In Our Image*, 180.
71. Linn, *The Philippine War*.
72. Coffman, "Batson of the Philippine Scouts," 68–70.
73. William K. Emerson, *Encyclopedia of United States Army Insignia and Uniforms* (Norman, OK: University of Oklahoma Press, 1996).
74. Vic Hurley, *Jungle Patrol, The Story of the Philippine Constabulary (1901–1936)* (New York: Cerberus Corp., 2011); Two Filipino officers began serving right away: Jose Velasquez and Felix Liorente.
75. James P. Ongkili, *Nation-building in Malaysia 1946–1974* (Oxford: Oxford University Press, 1985), 79.
76. Coffman, "Batson of the Philippine Scouts," 72.
77. The official record is unclear as to how many of those killed at Bud Bagsak were avowed combatants and how many were, noncombatants, women, and children and even how many of these were dressed like combatants or used as human shields. What is clear from the historical record is that a great many Moro's died at Bud Dajo.

78. 32 CFR 575.3 - Appointments; Sources of Nominations, n.d., https://www.law.cornell.edu/cfr/text/32/575.3.
79. Bernard Norling and Robert Lapham, *Lapham's Raiders* (Lexington: University Press of Kentucky, 1996).
80. Karnow, *In Our Image*, 340.
81. Agoncillo, *Malolos*; Karnow, *In Our Image*, 342; Benedict Kerkvliet, *The Huk Rebellion: A Case Study of Peasant Revolt in the Philippines* (London: University of California Press, 1977), 267; Michael McClintock, "Toward a New Counterinsurgency: Philippines, Laos, and Vietnam," in *Instruments of Statecraft: US Guerrilla Warfare. Counter-insurgency, and Counter-terrorism, 1940–1990* (New York: Pantheon Books, 1992).
82. McClintock, "Toward a New Counterinsurgency."
83. $2 billion in 1950, www.in2013dollars.com.
84. Karnow, *In Our Image*, 345; McClintock, "Toward a New Counterinsurgency."
85. McClintock, "Toward a New Counterinsurgency."
86. Karnow, *In Our Image*, 351.
87. Ibid., 354.
88. 1913–present. With the aforementioned exception of World War II.
89. Karnow, *In Our Image*, 200–201; Stuart Miller, *Benevolent Assimilation*, 3; Wolff, xi.
90. Karnow, *In Our Image*, 128 and 174.
91. Karnow, *In Our Image*, 170.
92. Karnow, *In Our Image*; and Stuart Miller, *Benevolent Assimilation*.
93. Karnow, *In Our Image*, 170.
94. Karnow, *In Our Image*.
95. Francis Fukuyama, *Origins of Political Order: From PreHuman Times to the French Revolution*, (New York, NY: Farrar, Straus and Giroux, 2011), 14.

CHAPTER 3

THE STRATEGIC RENTIER STATE

IRAQ

Around 4:30 p.m. on April 9, 2003, US Marines of Regimental Combat Team 7 (RCT 7) entered Fidros Square in Baghdad. They were led by M1A1 Abrams tanks and amphibious armored personnel carriers and were completely unopposed by Iraqi military forces. For the previous twenty days, the US Marines had fought through contact with Iraqi forces of varying degrees of intensity. They expected the last few blocks to the city's center to be the most difficult and bloody.[1]

Yet, upon entry into Fidros Square, they failed to encounter any Iraqi soldiers defending from complex interlocking urban fortifications that they had anticipated. Instead, the Marines found Iraqi civilians like Kadhim Sharif Hassan Al-Jabbouri, a local mechanic, slamming the base of a Saddam Hussein statue with a sledgehammer, while other Iraqis tried to pull the statue down with ropes and chains.[2] Amazed with the ease with which they had entered the city center of Baghdad, the US Marines then turned their hands to aiding Iraqis in pulling down Saddam's statue. For all intents and purposes, this signified the fall of Baghdad and the end of major conventional combat operations.

In 2010, after just shy of seven years of intervention, Iraq held parliamentary elections. By August 18, 2010, the last US combat brigade departed Iraq. On December 18, 2011, the last US troops left Iraqi soil. Then, on December 12, 2011, President Barack Obama and Prime Minister Al Maliki celebrated a joint press conference to announce the end of the Iraq War.

By July 2014, less than four years after the last American troops left Iraq in 2010, a relatively small handful of ISIS soldiers seized Fallujah, Ramadi, Baji, and Tikrit. They had also routed the two Iraqi infantry divisions, consisting of 30,000 Iraqi soldiers, and seized the city of Mosul with more than 650,000 residents. At the time, Iraqi forces altogether numbered nearly 300,000 active duty soldiers and over 500,000 reservists. Iraq possessed nearly 300 M1A1 modern main battle tanks, more than 100 self-propelled artillery pieces and 130 pieces of towed artillery, almost 60 multiple launch rocket systems, 14 attack helicopters, and more than 150 assault support helicopters.[3] In contrast, the ISIS forces arrived in Mosul in piecemeal fashion carried by pickup trucks with only around 1,500 fighters, (21,000–200,000 troops organization wide), no major armor or artillery assets, and while also simultaneously fighting on two to three fronts in Syria and Iraq.[4]

What accounts for the collapse of US-supported Iraqi security forces numbering as much as 30,000 in Mosul in the face of a 1,500 strong, ragtag militia in just four days? At the time, the Iraqi forces in and around Mosul possessed an approximate 30:1 advantage in manpower,[5] and they were defending from within complex urban terrain and supported by armor and artillery. Conversely, how did it take Iraqi forces three years to retake the same city back with nine months of bloody fighting? What accounts for the apparent failure of the US counterinsurgency and reconstruction programs in Iraq?[6] This was a $820 billion program that came at the cost of the lives of 4,474 US servicemembers and lasted 8 years, 9 months, and 3 days.[7] Lastly, how did the US produce a regime whose corruption index is ranked at 165 of 175 countries, with only ten other countries considered more corrupt than itself?[8]

EXAMINING THE CASE OF IRAQ

US Army General Tommy Franks was the commander of the US Central Command and of the US-led coalition which seized all of Iraq in little over three weeks. Because of the ensuing anarchy after the US invasion in 2003, an argument has arisen that General Franks did not really plan the stability and security phase, or "Phase IV" operations. However, as a 2008 Rand study notes that "several US government organizations, particularly the Office of the Secretary of Defense (OSD), the State Department (USDOS), the US Agency for International Development (USAID), and the National Security Council (NSC) conducted separate studies of postwar possibilities."[9] The problem, therefore, was not a paucity of thought about the challenges of post-Saddam Iraq. Rather, many of the complications that evolved after the invasion were the result of flawed planning assumptions; namely, that the Iraqi security forces would surrender en masse, that the Iraqi security forces, especially the Iraqi police, could be relied upon to help secure Iraq after the invasion, and the Iraqi government and Iraqi security forces would continue to govern and provide security to some degree until their leaders could be transitioned out slowly.

The first assumption—that Iraq would maintain a residual capacity to govern and provide internal security after being defeated—was a fundamental assumption made by US military planners. The presence of effective Iraqi security forces would greatly lower the number of US troops required to secure the country and would speed the withdrawal of US forces. However, this first of several key assumptions proved invalid. The Iraqi army, unlike European armies, did not surrender en masse but rather evaporated entirely once defeated. Additionally, this left an entire nation's allotment of weapons and munitions completely unsecured.

The second assumption—that the Iraqi security forces, and particularly the Iraqi police, could be relied upon to maintain the peace after invasion—was also invalidated. The Iraqi police had been corrupt and ineffective in the years preceding the US invasion in 2003.[10] Indeed, the uniformed police were only the most visible arm of the Iraqi security state that kept Saddam Hussein in power.[11] When the Iraqi security state was overthrown, the true day-to-day security apparatus, oppressive as it had been, was also overthrown. This left in charge the feckless and anemic face of the former Iraqi security state, the Iraqi police.[12] The Iraqi police were not prepared or equipped to handle the massive societal challenges of thirty-plus million people being released at one time from an oppressive security state. Iraq and the American forces in Iraq were faced with rampant, nationwide looting and anarchy, but they possessed insufficient forces to fill the void.

The third assumption— that the Iraqi government would continue to govern and the Iraqi army continue to provide security after the invasion—was invalidated not by force majeure but by coalition dictate.

It is not too strong a statement to argue that the scale of the insurgency that followed the invasion was largely one of the US's own manufacture. Flawed planning assumptions and the issuance of Coalition Provisional Authority Order #2 produced a pool of militarily trained and bitter future insurgents. The removal of the only effective security force and bureaucrats, as well as the provision of ample weapons and munitions to insurgents set the conditions for a nascent but well-equipped insurgency.

On May 6, 2003, L. Paul Bremmer, a former foreign service officer who had served twenty years with the US Department of State before retiring and going into the private sector in 1986, was appointed as the Presidential Envoy to Iraq. Bremmer arrived in Iraq on May 12, 2003, and he took over from retired US Army Lieutenant General Jay Garner, the head of the Office for Reconstruction and Humanitarian Assistance (ORHA), and became responsible for the reconstruction of Iraq after the ground offensive. Garner had been kept out of much of the direct planning

The Strategic Rentier State

for the invasion by General Tommy Franks and was replaced after less than sixty days by Bremmer. Mr. Bremmer led the Coalition Provisional Authority and was charged with the occupation and reconstruction of Iraq under UN Security Council Resolution (UNSCR) 1483. UNSCR 1483 authorized the occupation of Iraq by the US and the UK. It also required these two countries to reestablish stability and security and transition power back to the Iraqi people as expeditiously as possible.[13]

Bremmer issued Coalition Provisional Authority Order #2 on May 23, 2003. This directive ordered the mass de-Baathification of the Iraqi government and dissolution of Iraqi security forces.[14] This left Iraq and the Coalition Provisional Authority with more than 250,000 unemployed former soldiers. These quarter million disgruntled former soldiers were joined by 50,000 criminals, who had been released by Saddam before the invasion.[15] To compound the matter further, these criminals and former soldiers had access to massive quantities of munitions and weapons bypassed and left unsecured by US forces blitzing to Baghdad. Jay Garner and Tommy Franks had specifically advised against mass de-Baathification and dissolution of the Iraqi military.[16] Therefore, it is not too strong a statement to argue that the scale of the insurgency that followed the invasion was largely one of the US's own creation.

As a 2008 Rand report argued, the US troops were in Iraq in sufficient numbers to be seen as occupiers but in insufficient numbers to actually secure the country.[17] By the end of the invasion, US units were dispersed across Iraq in company-size defensive positions. In May and June 2003, these company-sized elements began to be pulled back into larger bases that required less force protection. Nevertheless, the "lines of communication" or LOCs necessary to supply these massive bases from Kuwait had to be secured. As violence and looting became rampant, US units found it difficult to do much more than secure the thousands of miles of LOCs necessary simply to maintain their own presence. US leaders were also reluctant to deploy US combat formations in major urban centers across Iraq. They feared inciting Muslim tensions due to the presence of Western military forces occupying major Muslim population centers.[18] As such, the US military leadership initially restricted US forces to massive bases and conducted surgical decapitation strikes and major offensive operations to defeat the burgeoning Iraqi insurgency.

COIN Methods Observed-Enemy-centric COIN

From 2003 until the start of the surge in later 2006, enemy-centric COIN methods were present and weighted, but they were not effective. Decapitation strikes are a common tool in irregular warfare and COIN. In Iraq, the pursuit of Saddam Hussein, former fedayeen, former Baathist leadership, and insurgent leadership were designed to decapitate insurgent leadership and end the Iraqi insurgency. Of the original fifty-five high-value individuals identified in 2003, only ten would remain unaccounted for by the end of the US's direct combat involvement in 2010.[19] Even with Saddam's death in 2006 and the death or capture of most of the high-value individuals on the US targeting list, violence continued to increase 600% from a 2003 baseline (see figure 3).[20]

By 2004, it became clear that the US was no longer dealing exclusively with former regime dead-enders. Instead, the US-led coalition faced an array of sectarian insurgent groups. As a result, the US shifted its surgical decapitation strikes to target leadership among Shiite and Sunni insurgent groups. The US employed different methods for targeting Sunni and Shia leadership. The US generally employed capture-kill raids to neutralize Sunni leadership. Perhaps the most famous of these strikes was the pursuit and killing of Abu Musab Al Zarqawi, the Al Qaeda leader in Iraq who had served in Afghanistan at the end of the Soviet War in 1989, then again from 1999 to 2001 where he fought with the Taliban against the US. He was tracked relentlessly and subsequently killed by a US precision guided bomb on June 8, 2006. However, this did little to curb the violence in Iraq from 2006 to 2008 (see figure 3).

The Strategic Rentier State 81

Shia militia leader Moqtada Al Sadr was handled very differently by the US-led coalition compared to Sunni insurgent leadership. Al Sadr, as the leader of the Jayish Al-Mahdi (JAM), or the Mahdi Army, began to target US forces because of the coalition's closing of his newspaper *Al-Hawza* in the spring of 2004. The Coalition Provisional Authority sought to marginalize Al Sadr by labeling him an outlaw. However, the coalition refused to seriously attempt to arrest or kill him. US forces feared upsetting the majority Shia who venerated Sadr's martyred father, Mohammad.[21] As a result, US forces began to target the lower-level JAM members instead of its senior leadership like Al Sadr. With the Battle of An Najaf in 2004, the JAM forces absorbed a costly blow from the US operations. The battle caused Al Sadr to rein in his militia to a degree, but it did not remove the JAM or Al Sadr. The violence only continued to increase.

After the US invasion, the principal security focus for US forces was to secure the LOCs between major bases and to root out major enemy strongholds through massive clearing operations.[22] US forces tried in vain to secure their LOCs by establishing interconnected observation posts every few hundred meters. However, given the thousands of miles of Iraqi roads, US forces could only cover small portions of these major routes. Second, as the insurgents began attacking small coalition formations, the coalition began to regulate the size of units that could leave the security of the bases. By late 2003, the coalition began directing that units could not "leave the wire"[23] in patrols smaller than four vehicle convoys or dismounted patrols smaller than eight troops.[24] This focus on reducing risk by employing larger formations did provide the enemy with harder targets, but it failed for the most part to significantly reduce violence against coalition forces.

Operation Al Fajr in Fallujah, Iraq, is perhaps the preeminent example of a major American clearing operation of the Iraq War. In early 2004, the US Army in Fallujah evacuated all soldiers from Fallujah due to violent protests that resulted in as many as fifteen dead Iraqi citizens.

Thereafter, US soldiers only conducted raids and patrols into the city. These operations resulted in skirmishes which did little to diminish the insurgency. By March 2004, responsibility for Al Anbar province was transferred to the 1st Marine Expeditionary Force (I MEF). Before the end of the first month of responsibility for Fallujah and Al Anbar province, the US Marines were compelled by the Bush Administration to react precipitously to the ambush of Blackwater contractors resulting in Operation Vigilant Resolve. Receiving instructions from the highest levels of the Bush Administration, and against their advice, Marine generals assembled an ill-prepared urban offensive. The assault stalled when the US Marines lacked sufficient combat power to cordon off the city and clear through the complex urban defense. The US effort also stalled psychologically as insurgent propaganda dominated all the narratives coming out of the battle. In order to save face, US commanders experimented with the Fallujah Brigade, an ad hoc unit led by former insurgents and Baathists, to provide security for Fallujah. Once the Marines withdrew, this ad hoc force dissipated.

This 1st Battle of Fallujah became a pyrrhic victory for the insurgents. It was a clear tactical and propaganda victory for the insurgents. However, it also made the US effort during the 2nd Battle of Fallujah that much more effective. The Sunni insurgents had provided the US a major, static target upon which the US could bring its massive firepower advantage to bear. After the 1st Battle of Fallujah, US commanders immediately began preparing to retake Fallujah. They began assembling sufficient force to clear and hold a major urban center of 275,000 Iraqi citizens. By November 7, 2004, the US amassed assault and cordon forces of approximately 9,000 US and Iraqi troops. These were organized into two Regimental Combat Teams (RCT 1 and RCT 7). The assault force was supported by tanks, armored personnel carriers, close air support, and combat engineering and breaching assets for clearing through and/or avoiding major insurgent obstacles. On November 7, 2004, Operation Al Fajr (the Dawn)[25] commenced. Two days before Christmas 2004, the assault force had cleared Fallujah. The operation killed an estimated 1,350–

2,000 insurgents, captured around 1,200–1,500 suspected insurgents, and routed 3,500–4,000 more.[26] While costing seventy American and 107 coalition lives total, the operation was a tactical success as well as a moral victory.[27] Still, despite the short-term victory, Fallujah would once again return to high levels of insurgent control by 2006. The situation would eventually become so severe that Fallujah would require another major population-centric COIN operation to pacify it in 2007–2008.[28]

Fallujah was by no means the only major battle/clearing operation during this time. After a firefight between the JAM and US Marines from 1st Battalion, 4th Marines on August 5, 2004, a major pitched battle ensued in the vicinity of the Imam Ali Mosque. Similar to Fallujah, Shia insurgents provided US forces with static targets for American fire support assets. Insurgents were cordoned, targeted by close air support and artillery, and closed upon by infantry. The JAM was severely punished as a result of the Battle of An Najaf.

As a result of the perceived successes of decapitation strikes and major named operations in Fallujah and An Najaf, the US continued major clearing operations from 2004–2006. Major clearing operations like the Battle of Samarra September 2004, the Battle of Abu Ghraib April 2005, the Battle of Al Qaim May 2005, the Battle of Tal Afar September 2005, Operation Steel Curtain November 2005, and the Battle of Ramadi June 2006 all sought major pitched battles and results similar to Fallujah and An Najaf.

However, after Fallujah and An Najaf in 2004, insurgents no longer provided large static targets for US firepower. From 2005–2006, insurgents adapted and side-stepped major offensive operations in favor of IED strikes, ambushes, sniping, targeted killings of Coalition supporters, and small-scale raids. Nevertheless, the US continued to use fruitless major clearing operations and short duration patrolling of LOCs that allowed the insurgents nearly unfettered freedom of maneuver only a short distance from these LOCs.

From 2004 to 2007, violence exploded nearly 700% (see figure 3). From just the start of 2004 to the summer of that same year, SIGACTS increased over 150% and remained steady until the early fall. By the fall of 2004, sectarian violence began to take shape as local militias began to fill the void emptied by the US and Iraqi security forces. Violence swelled, increasing to almost 200% more than in early 2004, and remained at almost 600 SIGACTS per month until late summer 2005 (figure 3). Following the constitutional referendum in October 2005 and the parliamentary elections of December 2005, violence continued to increase to nearly 300% from the 2004 baseline. By February 2006, with the bombing of the Shiite Golden Mosque in Samarra, violence further escalated 400% from 2004 levels as a result of the de facto civil war between Sunnis and Shia—even as both sides continued to attack American forces. This was the environment confronting the Coalition Provisional Authority and Combined Joint Task Force 7/Multinational Force Iraq just prior to the surge.

The Surge—Population-centric COIN

From late 2006 until 2010, population-centric COIN methodologies were present, weighted, and effective elements of US operations in Iraq during this time. The surge is an unfortunate name for the successful population-centric COIN operations that turned around the Iraq War from 2006–2009. The Iraq surge, led by General David Petraeus from 2006–2008, represented the penultimate application of contemporary COIN doctrine and methods. The surge's strategy integrated classical population-centric COIN theory with an impressive amalgam of contemporary scholarship and praxis.[29] The Iraq surge was executed with nearly unconstrained economic and combat support resources by the most educated American military force the US has yet possessed. Even the surge's initial opponents were obliged to acknowledge its success.[30] However, the boon of this success was short-lived after the US withdrew.

Figure 3. Violent Significant Actions (SIGACTS) 2001–2009.

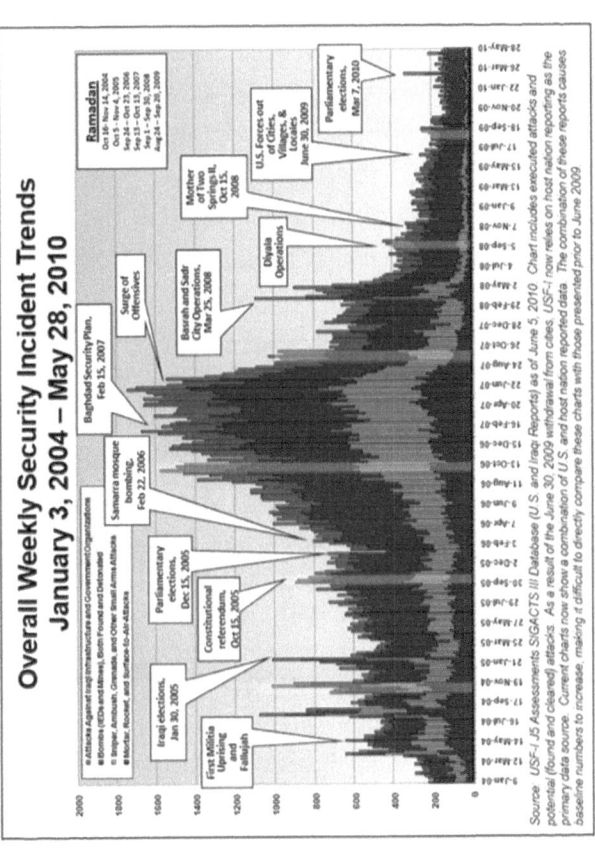

Source. Adapted from *Measuring Stability and Security in Iraq-Report to Congress*, (p.27), 2010, Arlington, Virginia: US Department of Defense.

The surge had less to do with the number of US troops "surged" into Iraq and more to do with a sea change in how the war was being fought. Indeed, even assuming the most extreme increase in US troop numbers, this accounts for only 16,700 more troops. A significant portion of this increase came not from adding more troops but by extending troops already in theater. Examining only the increase in US force size fails to acknowledge the near-simultaneous reduction in troop levels from other Coalition countries. This will be discussed in greater detail under Troop Ratios. Nevertheless, the increase in troop levels cannot account for the dramatic changes brought on by the surge.

One of the most important COIN adaptations of the Iraq War was the 2006 Field Manual 3-24 (FM 3-24), *Counterinsurgency*. The implementation of FM 3-24 (2006) is credited with both guiding COIN strategy under Petraeus and the initial successes in Iraq and Afghanistan since 2007.[31] The FM 3-24 (2006) is peculiar not only because of its reinterpretation of classical COIN theory but also because of its provenance from the intersection of military and academic professionals.

The focus of Petraeus's new strategy would be protecting the population and separating the insurgents from their support among the population.[32] The most common historical means of physically separating the populace and insurgency during the US Civil War, the US-Philippine War, and the Vietnam War was to relocate the population itself into more defendable locations. The US forces could then protect the populace and control the movement of the insurgents among the populace. By isolating the insurgency, the COIN force could thereby literally starve the insurgency of its logistical, intelligence, and recruiting support.[33] However, in Iraq, the mass relocation of hundreds of thousands of residents from densely populated urban areas was not a realistic option.

Rather than relocating hundreds of thousands of civilians, US forces in Iraq employed a combination of static and dynamic means to physically and morally isolate the insurgents from the populace. Using the concept of a "gated community," US forces erected static barriers around all the major population centers. To the barriers, the US forces added biometric identification systems to help identify innocent Iraqi civilians from insurgents. To solve the problem of persistent observation around the communities, US forces transitioned UAVs, snipers, and cameras away from watching LOCs to monitoring access into, inside, and around population centers. To these physical assets, US forces added dynamic ground operations to isolate the insurgency as well.

These dynamic operations included patrolling, random vehicle checkpoints, and entry control points to scan and search everyone entering a population center. Once inside a population center, US troops further separated insurgents from the populace by targeting them with a spectrum of human, signals, and ground intelligence operations. The intelligence produced from these operations could then be followed up by targeted raids to capture or kill these insurgents, while seeking to limit collateral damage. These capture-kill operations also evolved with the publication of FM 3-24 (2006). Operations gradually moved away from aggressive "cordon and search" raids to more permissive "cordon and knock" operations that requested permission to enter before conducting forcible entry. These newer cordon operations began to rely increasingly on Iraqi forces and American female service members to conduct the searches. The Iraqis and US female units, like those in the Lioness Program and Female Engagement Teams, had access to Iraqi men and women that American males could not gain.

Lastly, US forces could no longer afford to dwell in massive US bases and physically isolate the insurgents from the populace. Under General Petraeus, US forces gradually began to accept the greater risks associated with distributing forces in increasingly smaller sizes over progressively greater areas. The full extent of this was reached by mid-2007. It was not uncommon to have tiny US forces of six to twelve US troops and a handful of Iraqi police or soldiers in Joint Security Stations (JSSs) spread out in densely populated urban areas. This allowed these troops to develop a beat cop mentality where they would get to know the streets and residents in their neighborhood to determine who or what did or did not belong. These efforts to physically isolate the insurgency from the population were complimented by moral isolation of the insurgents as well.

FM 3-24 (2006) describes insurgency and counterinsurgency as battles for legitimacy between insurgents and counterinsurgents among the population. Both sides seek to isolate the population morally as well as physically from the other. In support of this effort, US forces under Petraeus developed lines of effort/lines of operation[34] to synchronize their efforts to legitimize US and Iraqi government operations while also delegitimizing insurgent operations. These lines of effort/lines of operation included security, governance, rule of law, essential services, economic development, and security force development.[35]

Security was devoted to the defeat of the insurgents and their physical isolation from the population. Governance and rule of law was executed through military assistance to governance operations whereby US officers would operate with local and provincial councils to advise them on rule of law and governance issues. Essential services such as sewage, water, electricity, and trash were also addressed. As Nagl notes, with respect to the Commanders Emergency Response Program or CERP funds, US commanders could readily pay for essential and immediate local Iraqi needs. Nagl notes that programs like CERP went a long way towards winning Iraqi hearts and minds to US forces.[36] The Iraqi government and security forces, however, were either unwilling or unable to continue these practices on their own after the US forces departed.[37]

Perhaps the most consequential adaptation in isolating and protecting the populace was the cooption of disaffected Sunni tribes to combat Sunni insurgents in Al Anbar. Al Sahwah, or The Awakening movement, began with Sheik Satar's Albu Risha tribe outside Ramadi, Iraq in 2006. The Awakening then found its way into central Ramadi as a result of support from the Al-Alwani tribe in northern Ramadi. The Al Anbar Awakening exploded principally as a response to the atrocities against Iraqi citizens inflicted by Sunni and Shiite insurgents. The severity of these atrocities by Sunni insurgents cannot be overstated. It was not uncommon for US forces to be taken to Sunni insurgent torture chambers by brutalized Iraqi citizens and Iraqi security forces after an area had been won back from insurgents. By mid-2006, Al Qaeda in Iraq and Islamic State excesses had become so egregious that former suspected insurgent leaders like Sheik Sattar encouraged tribal members to work with US forces. From 2006 to 2008, the Islamic State of Iraq gradually lost military ground to US, Iraqi, and local Sunni tribal forces.

The isolation and protection of the Sunni populace, the securing of Baghdad's neighborhoods, and the defeat of Al Qaeda in Iraq opened the door to Sunni participation in the Iraqi government.[38] By the time of Zarqawi's death in June 2006, violent SIGACTS numbered over 1,000 per month just as the surged US forces began to arrive. Just before the start of the surge offensives began in mid-2007, the violent SIGACT rate was the highest of the entire war at just under 1,600 per month. Impressively, in less than six months the number of SIGACTS would be more than halved and within a year's time the number would drop to below 2004 levels (figure 3). In less than a year and a half, SIGACTS would be further reduced to nationwide levels not seen since the start of the invasion. However, the reduction of SIGACTs was not the only positive outcome. By the provincial elections of 2009, many Sunnis had finally been brought into the democratic process. Moreover, civilian deaths declined by 45%, and ethno-sectarian deaths were down a further 55%.[39]

Troop Ratios
In contrast to the Philippine case, US and coalition troop ratios in Iraq did fall well within the range of 1:50 and 1:91 counterinsurgents to populace. Yet this appears to have had no positive impact on SLAW. During the height of the Iraq War, the coalition possessed a 1:70 counterinsurgent-to-population ratio. This is significantly more than the 1:50 recommended by Quinlivan but well within tolerances of McGrath. However, if the same standard that Linn and Krepinevich argue for regarding foxhole strength is relied upon, then the number of combat effective personnel

of the coalition force must be reduced to 60%.[40] This results in 277,020 in FY2006[41] or a 1:121 counterinsurgent-to-population ratio.[42] In FY2008, there were an estimated 287,040 total combat-effective coalition troops, which lowered the ratio to 1:116 counterinsurgents to population.[43] Though both of these numbers are outside even McGrath and Brown's tolerances, they are substantially better than the combat-effective ratio of US and Filipino counterinsurgents to population in the Philippines from 1898 to 1913.

Troop ratios also fall short of explaining the failures of the Iraqi government and security forces in 2014. Before the surge, with Iraq's 33.4 million citizens and the maximum number of US boots on the ground of 141,100 (FY2006), this provides a ratio of 1:237 counterinsurgents to populace.[44] Further, if the 15,000 other coalition members are added to the total force along with the Iraqi security force members (227,600), then there were a total of 383,700 coalition troops on the ground just before the surge.[45] Lastly, the other 78,000 contractors increase the total counterinsurgent force to approximately 461,700 troops. This results in a 1:72[46] ratio of counterinsurgents to populace for the country or roughly 2.74[47] troops per square mile.[48]

The difference between FY2006 and the peak of FY2008 is only 16,700 troops, or one additional US soldier or Marine for every 2,000 Iraqi citizens.[49] Moreover, the number of non-US coalition forces[50] actually dropped from a high of between 18,000–23,000 troops just before the surge down to only 15,000 in early 2006. Therefore, with the additional 16,700 US forces and the loss of 3,000–8,000 Coalition troops by FY2008, the total coalition force was 478,400 for a counterinsurgent-to-population ratio of 1:70.[51]

Table 2. Average Monthly US Troop Levels in Iraq War.

Average Annual Troop Level	FY2002	FY2003	FY2004	FY2005	FY2006	FY2007	FY2008	% Change 04-08
Boots on the Ground	0	67,700	130,600	143,800	141,100	148,300	157,800	21%
Operations Report	0	128,60	196,900	184,400	181,300	191,200	193,100	-2%
Combat Pay Estimate	0	123,200	205,700	202,000	197,400	231,000	236,800	15%
Average Strength with CRS Allocation	0	216,500	211,600	230,500	234,100	243,200	251,100	19%
Location Report	NR	NR	NR	NR	229,587	242,531	252,126	NR

Source: Joint Staff, Joint Chiefs of Staff, "Boots on the Ground" (BOG) reports; Joint Staff, "Operations Report" Defense Finance Accounting Service (DFAS), "Supplemental & Cost of War Execution Reports;" Defense Manpower Data Center (DMDC). DRS 21198, "Average Number of Members Deployed on any given day by Service Component and Month/Year," January 2009; DMDC, DRS 11280, "Modified Location Country Report," December 2008.

The increase in US boots on the ground is an unsatisfactory explanation for the success of the surge in the short term. The counterinsurgents-to-population ratio only changed from a 1:72 to 1:70 by FY2008 when the significant decrease in violence occurred. In other words, a coalition soldier's responsibility only decreased from 72 citizens to 70. This delta of the ratio is an unsatisfying answer to the question of why the surge worked in the short term and why US efforts failed in the long term.

Enduring Insurgent Sanctuaries
Whereas the US was very successful at isolating population centers during the surge, at no time did the US completely foreclose on insurgent sanctuaries. Insurgent sanctuaries, training centers, and infiltration routes for Iraqi and foreign fighters existed in Iran and Syria throughout the US-led portion of the Iraq war.[52] Though these were an enduring concern for military planners throughout the war, their presence was not sufficient to prevent the success of the surge.

Indeed, a key element of population-centric COIN is separating the insurgents from their popular support. In some ways, these insurgent sanctuaries worked in favor of the coalition forces by voluntarily separating insurgents from population centers in Iraq. This is not to say that sanctuaries are not a legitimate concern in most COIN operations; adversary sanctuaries, for example, were used in Vietnam to train and equip Phase III guerilla forces.[53]

Sanctuaries in Syria were used to allow ISIS to prepare for its 2013–2014 invasion of Iraq. However, Syria was hardly a sanctuary in the sense that ISIS used it as an unmolested space within which to train and prepare. ISIS had to contend not only with moderate Syrian forces backed by the US but also Syrian regime forces backed by Iran and Russia. Still, despite its best efforts, the US never foreclosed on ISIS sanctuaries in Syria during the US intervention. Similarly, Shiite insurgents traveled freely into and out of Iraq from Iran and received training there.[54]

Continued Support After Withdrawal

Levels of US military and economic aid to Iraq from 2011 to 2014 also fail to explain the failure of the Iraqi government and security forces in the face of pressure by ISIS from 2014 to 2017. However, the absence of US combat support and advisors does help explain this failure. Whereas US economic assistance dropped initially from $811 million in 2012 to $383 million in 2013, that assistance has remained fairly steady ever since (see table 3). Moreover, Iraqi GDP growth has actually grown 10% since the US's withdrawal and in 2013–2014 was at its highest in Iraq's history.[55]

Table 3. Total US Economic and Military Assistance, 2012–2015.

	2012	2013	2014	2015
Total US Economic Assistance	$811.2m	$383.7m	$392.0m	$394.4m
Total US Economic Assistance	$1.158b	$62.2m	$22.1m	$180.3m
GDP	$218b	$234.6b	$228.7b	$180.7b

m-million b-billion

Source. Adapted from US Overseas Loans and Grants: Obligations and Loan Authorizations, July 1, 1945–September 30, 2015, World Bank 2017, http://www.trading economics.com/Iraq/gdp.

The biggest drop in the Iraqi government's revenues came from US military assistance, which dropped from around $1.2 billion in 2012 to just $22.1 million in 2014.[56] This reduction in military assistance apparently had little to no effect on the Government of Iraq's overall budget. According to Luay Al-Khatteeb of the Brookings Doha Center, "Iraq's federal budget has increased to five times its size between 2004 and 2015. No matter how much oil revenue enters the treasury, budgets always have a deficit of around 20 percent, while actual spending always amounts to 70 percent or more. This leaves less than 30 percent for investment and development."[57] Iraq is producing more oil today than it has since the start of the Iraq War in 2003. Despite a drop in global oil prices, the Government of Iraq still receives 90–95% of its revenues from its sale of oil.[58] Therefore, it does not seem probable that the reduction in

US economic support was the cause of the failures of the Iraqi government and security force in 2014.

Additionally, the lack of continued military economic support fails to explain the Iraqi government and security force failure in 2014. The $1.6 billion in US military assistance provided after 2014 was not used to create a military capacity that did not previously exist. Rather, it was used to retrain and re-equip a military that had already been trained and equipped at a cost of more than $20 billion over nine years.[59] Many of the key shortfalls and issues that US advisors have been trying to address since 2014 are the very same ones US advisors noted in 2003–2010: intelligence, logistics, corruption, and leadership.[60] In many ways, the money, time, and effort spent by the US in Iraq since 2014 has been to get the Iraqi militarily back to where it was when it was first overrun by ISIS in 2014.

Based on reports from Iraqi troops on the front lines in Mosul, the issue was not that the Iraqi Army lacked resources. Rather, the problem appears to have been the manner in which the Iraqi army used the forces and resources it had. One Iraqi commander related that he had requested tank support and got a Soviet-era antique T54/55 tank with an untrained driver instead of a US-supplied M1A1 modern main battle tank and American trained driver. The commander claimed that this was because he was not willing to send along a $2,000 bribe.[61] Even when Iraqi troops were willing and well led, it appears that the lack of the most basic logistical operations hampered efforts. American advisors observing operations in Iraq in 2016 were frustrated watching entire assaults halted because Iraqi troops could not get sufficient water to operate in the stifling heat. This is an obvious requirement for desert operations and particularly so considering the Iraqis were operating within their own country. Lastly, even the best-led Iraqi units "have struggled to hold on to a series of hamlets, while soldiers admit that they have largely relied on US air support to advance. The situation was similar in larger victories, from Ramadi to northern Sinjar, where US-led strikes flattened the way for ground forces."[62] Therefore, the failure of the Iraqi government and security forces in 2014 cannot be attributed to a failure of economic support. It does, however, point to the fact that the US had trained the Iraqi forces to rely on US offensive air support and supervision.

Governance and Security Force Development

Degrees of Embeddedness in Governance Development

From 2003 to 2010, the US chiefly relied on institution-influencing strategies to develop the Iraqi government. The exception to this larger strategy of low embeddedness was the Coalition Provisional Authority's control of Iraq from April 2003–June 2004. Nevertheless, for the bulk of the intervention, the primary means the US employed to develop governance in Iraq was through military assistance to governance and ministerial advisor teams.

Larry Diamond, senior advisor to the Coalition Provisional Authority in 2003–2004, offers that, "the CPA presided over an ambitious effort to promote pluralist democracy in Iraq..."[63] However, this new democracy was not the bottom-up, combined institutional adaptation shared by American and Filipino governments in the Philippines 1898–1913. Rather, it was a top-down, rushed democratic pretense designed to impose exclusively American progressive adaptations that emphasized national divisions and distinctions rather than unity. According to Saad Jawad of the London School of Economics, the US-led constitutional effort liberalized an as yet undemocratic society well beyond the tolerances for progressivism that even an established democracy like the US could have handled.[64]

The US spent the better part of eight years in the lead for the military effort to secure Iraq. In contrast, it only spent fourteen months artificially imposing a completely alien form of government devoid of constituents from the top down. The coalition imposed this alien form of representative government during a period of the Iraq War when the deteriorating security situation prevented even the most basic connections between Iraqi politicians and Iraqi citizens.

The period of 2003–2004 was so dangerous that the Coalition Provisional Authority could not spend the bulk of the $18.6 billion allocated for Iraqi reconstruction because it could not leave the Green Zone safely to do it. The Iraqi security situation was so unstable that in 2004, "insecurity drove the political occupation into a physical and psychological bunker. Already separated from Iraqis by the formidable security around the three-square-mile 'Green Zone' ...coalition officials began to travel less and less with every passing month. By the early spring of this year, foreign officials and contractors could no longer safely move around the country without an armored car and a well-armed escort."[65]

In 2003, British officials in Basra requested permission from the coalition to hold direct local elections. This would have allowed citizens in the British area of responsibility to elect their own local governing councils. These local governments would have facilitated the creation of grassroots constituencies to represent local citizens on the provincial and national levels as these structures began to take shape. Throughout Iraq's provinces, and in most cities and towns, local coalition military commanders worked with Iraqis to form representative councils. Yet, the Coalition Provisional Authority repudiated any forms of direct local elections for fear that these would undermine the Coalition Provisional Authority's argument that direct elections could not be organized so soon.[66] This rejection of local representation would remain a consistent theme by the Coalition Provisional Authority and later the Iraqi government, even as late as 2008. By then, councils who had helped force Al Qaeda out of places like Ramadi would be marginalized by less representative provincial and national bodies.[67]

Instead of building Iraqi governance from the local to the national level, the Coalition Provisional Authority installed a figurehead Iraqi governing body to act as a *kafil* and lend legitimacy to its efforts. According to Giacomo Luciani and Hazem Beblawi, a *kafil* is a local sponsor in a rentier state who is employed by a foreign entity to legitimate its independent operations in another state.[68] To this end, the Coalition Provisional Authority created the Iraqi Governing Council in July 2003. The Iraqi Governing Council was a multiethnic organization dominated by former Iraqi expats and Kurdish representatives.[69]

The twenty-five members of the Iraqi Governing Council were selected by quotas related to their ethnicity and their loyalty to the US. On its surface, appointment based on ethnic quotas seems logical in terms of their ability to represent previously marginalized ethnic constituencies. However, according to Jawad, this was the first time in Iraq that representation was based solely on a sectarian and ethnic basis. Also, the Iraqi Governing Council lacked any real sense of representativeness. Apart from the Kurdish leaders, "only five members of the Iraqi Governing Council were living in Iraq before 2003. Sixty-five per cent of the Iraqi Governing Council also held other nationalities."[70] Not only were these former expats out of touch with their supposed constituents, the violent security situation at this time prevented any meaningful contact with their citizens. Indeed, most Iraqis never saw any of their council members.[71]

One of the first steps towards transitioning Iraq back to self-rule, as prescribed by UNSCR 1483, was to establish a basic law under which a new constitution could be created. Under the supervision and direction of the Coalition Provisional Authority, the Iraqi Governing Council was given until February 28, 2004, to develop an interim constitution. The Transitional Administrative Law was received by some like Jawad as the work of a rushed, unelected body that had no constituents outside the Green Zone. The Transitional Administrative Law called for an unelected Iraqi Interim Government to be created by the unelected Iraqi Governing Council, based on consultation with the UN mission to Iraq. The document made strong provisions for individual rights and began to address the thorniest issue: deciding how to divide power within the government and the regions.

Jawad points out a unique distinguishing feature to the Transitional Administrative Law. "It is important to add here that constitutions drafted for disunited countries tend to concentrate on points of unity rather than division. For example, the US constitution, in an attempt to create a united country, stressed the importance of unity. The fact that 'sect' was mentioned in Iraq's constitution became a strong argument for those demanding an expansion of the quota system."[72] He has argued that the effort to create a unified central government was compromised by divisional and sectarian language built into the Transitional Administrative Law, which appeared to be relying less on the unifying language of the US Constitution and more on US twenty-first century micro-sectarian advocacy politics. In its rush to complete the Transitional Administrative Law without national debate and without any local consensus, the Iraqi Governing Council papered over genuine societal rifts that could not be legislated away. Local political groups denounced the document as an "unfair, unrepresentative, and undemocratic dictatorship of the minorities."[73] Nevertheless, under Bremmer's pressure, the Governing Council worked into the late hours of February 28 and signed the Transitional Administrative Law on March 1, 2003.

On June 1, 2004, the Iraqi Interim Government was formed in large measure from members of the former Iraqi Governing Council. During the month of June 2004, a new cabinet was selected. The former chair of the Iraqi Governing Council's security committee, Iyad Allawi, was chosen as prime minister, and Sheikh Ghazi al-Yawer was chosen as president.[74] The actual transfer of authority from the Coalition Provisional Authority to the Iraqi Interim Government took place on June 28, 2004. The US's official occupation of Iraq ended, and L. Paul Bremmer flew home hours afterward. The rushed transfer of authority was designed to steal the political initiative away from terrorists in Baghdad.[75] However, because the affair was largely hidden from public view and behind layers of American security, it did little to build support among the populace.

President George Bush relished the opportunity to declare that, "Fifteen months after the liberation of Iraq, and two days ahead of schedule, the world witnessed the arrival of a fully sovereign and free Iraq."[76] President Bush's claim of full Iraqi sovereignty hardly described the situation on the ground in 2004. The Iraqi Interim Government had no control over the massive US forces in Iraq. The Iraqi Interim Government did not possess an Iraqi security force with which to guarantee security. It had no method of extractive power to collect taxes to fund the government, and it was barred from making any long term, unilateral policy decisions.[77] This calls into question the utility of such a perfunctory and baseless political statement as proclaiming full Iraqi sovereignty.

On January 30, 2005, Iraqis voted for the transitional Iraqi parliament. In the largely Sunni Al Anbar province, turnout was very low. Though the election saw the ad hoc creation of more than 120 parties, less than 2% of Sunni Arabs participated.[78] Nevertheless, the election was a boon for Iraqi Shia, Kurdish Iraqis, and the US. The Shia, as expected, were the big winners. With the Sunnis out of the election, Shia parties were able to gain more seats than would have been possible otherwise. Moreover, the Kurds also benefitted from the Sunni boycott with more seats and their first meaningful representation in Iraq at a national level. Finally, the US was also a big winner, at least internationally.

Some of the successes Diamond identifies as wins during this period include support to emerging society organizations, imposition of gender and ethnic quotas in the legislature, training programs for political parties, and development of humanistic studies centers. In support of this ambitious project, financial assistance and technical support was provided to women's groups and think tanks to stimulate democratic participation. Perhaps one of the best examples of this uniquely American adaptation is the arbitrary establishment of a female legislative quota. The Transitional Administrative Law required that, "Every third name must be a woman's, to ensure that at least 25% of the seats in the assembly go to women."[79] Additionally, Diamond reports that millions of dollars were spent on creating a network of eighteen democracy centers in Iraq's primarily Shia south and central provinces. These were created to advance human rights and women's rights and assist Iraqis to "mobilize and organize" politically. This same financial and technical assistance was used to build a university for humanistic studies in Hillah, a Shia-dominated part of Iraq.

Saad Jawad interprets these efforts at mobilized pluralism differently. He argues that the US spent great energy ensuring the ability of human rights groups and minorities to mobilize, even as the political process continued without the participation of a massive ethnic group—the Sunnis. Jawad also observes that with the requirement for gender quotas in elections to the Iraqi legislature, the Coalition Provisional Authority tried to impose contemporary American micro-sectarian political advocacy before the country had a functioning state. Lastly, both Jawad and Diamond note that these radical democratic initiatives were carried out with little or no contact with average Iraqis. Due to the menacing security situation, Iraqi legislators and bureaucrats were perpetually sequestered behind American Green Zone security and unable to travel about freely.

February to October 2005 was devoted, in large measure, to the creation of an Iraqi Constitution and the conduct of a national referendum on this instrument. The Iraqi Constitution was drafted with minimal Sunni participation and approved via national referendum on October 15, 2005. Next, parliamentary elections, as described in the new Iraqi Constitution, were conducted on December 15, 2005. As a result, Nuri Al Maliki was seated as Prime Minister. Jalal Talabani, of the Kurdish Alliance, was seated as President. Shia Arabs make up over half of the population of Iraq, so, not surprisingly, out of the 275 seats in the National Assembly in 2005, more than 180 went to Shia political groups, or roughly 65%. However, the Kurdish population is only half that of the Sunnis or about 10–15% or the total populace, but they took 75 of the 275 seats available, or 27%.[80] Despite its apparent successes, Jawad takes issue with the haste and shortfalls of the 2005 Iraqi Constitution.

In 2005, there was a legitimate argument to be made that the failure of the Sunnis to participate in the elections resulted in a self-imposed disenfranchisement. However, there is another argument that benefits from hindsight which would militate against a democratic process that excludes a quarter of the population. There was little chance of fair, equitable elections and representation of the majority of the country where 25% of the electorate refused to participate. To make matters worse, this boycotting 25% had already demonstrated a willingness to achieve through violence what they could not achieve through democracy. Moreover, when Sunnis observed the election of even Shia insurgent leader Muqtada Al Sadr's party, especially given the liquidation of so many of their own insurgent leaders, there should have been little surprise that the Sunni insurgents would resort to violence.

Unlike the manner in which American-style governance was taught in the Philippines, in Iraq the responsibility was shifted to the Iraqis well before there was any demonstration of capacity. The administration of a completely novel form of government was taught on the job through the US Embassy's Ministerial Assistance Teams. According to the DoD's report to Congress in August 2006, "these teams, composed of civilian and military experts in governance and organizational development, mentor and train both the Iraqi ministers and their senior staffs..."[81] In addition to Ministerial Assistance Teams, US governance advisors constituted the Iraq Reconstruction Management Office in order to advise the Iraqi government on policy matters related to essential services and energy. However, these groups, the Ministerial Assistance Teams and the Iraq Reconstruction Management Office, also operated alongside the Iraqi government—cloistered in the Green Zone and alienated from their constituents.

US legal advisors helped rebuild the Iraqi legal system. However, their efforts were hampered by an overburdened prison system and the confusing authorities of the Iraqi legal system paralleling the US military detention system. Any pretense of Iraqi sovereignty was already delegitimized by the need for US forces to conduct lethal operations and detentions against Iraqi citizens on behalf of the Iraqi government. By July 2006, coalition forces held 12,388 security internees outside of the Iraqi legal system, who had to be transferred officially into the Iraqi system or released.[82] The Iraqi system lacked legal authority to hold these security internees outside of the judicial system, and the coalition could not hold on to them indefinitely either.

To compound the problem, the Iraqi judicial system in 2006 only had about half of the required 1,500 judges and only anticipated an increase to about 60% of those required by 2007.[83] With the insurgents deliberately targeting judges who were not protected inside the Green Zone, only twelve panels existed to handle more than 100 insurgency-related cases per month. Because of these stressors, US units throughout Iraq were forced to conduct massive detainee releases from 2007 through to the end of the war. This was mitigated somewhat by the fact that many of the detainees released from places like Camp Bucca were either innocent or had made professions of loyalty to the Iraqi government. However, because ISIS used places like Camp Bucca for recruiting and training jihadists, there were quite a few dangerous individuals released as well. Abu Bakr Al Baghdadi was one of these.[84]

Outside the Green Zone, the US operated four Provincial Reconstruction Teams (PRTs) in 2006 to assist in developing provincial level governance. These PRTs grew to twenty-five teams before the last PRT was disestablished in 2011. In Iraq, PRTs were "civilian-military organizations designed to operate in semi-permissive environments. PRTs were intended to achieve political objectives, counterterrorism and promote social and economic development."[85] The term "semi-permissive" reflects the difference between the PRTs and elements like the Ministerial Advisor Teams and Iraq Reconstruction Management Office. PRTs were able to operate in areas where some risk was possible. The first PRT was established in Mosul in late 2005, and, in Iraq the PRTs were constituted in nearly inverse fashion to the later Afghanistan PRTs. These PRTs were almost exclusively comprised of civilian personnel and led by a US Department of State Foreign Service Officer.[86] Later, in Afghanistan, the PRTs would be made up almost entirely of military personnel and a small contingent of civilian professionals. In Iraq, only a small fraction of the PRT was military, and the teams were tied to large US bases unless they could negotiate military transportation and protection. Therefore, while the PRTs in Iraq were closer to where the reconstruction efforts were actually taking place, they were only slightly more capable of direct engagement with the populace they served than US and Iraqi elements in the Green Zone.[87]

A key governance adaptation of the surge was the modification to ePRTs (embedded Provincial Reconstruction Teams). In early 2007, with the announcement of the surge, ten new ePRTs were established. These would be embedded within US brigade–sized elements as part of their staff rather than operating alongside combat forces. US military and civilian efforts were combined at the brigade level to advise Iraqi governance, rule of law, economic development, and essential services, thereby avoiding parallel, disjointed efforts. This combination was meant to address gaps that military and civilian officers would have likely experienced had they undertaken these efforts without each other. Military officers had little expertise or training to advise provincial governments. They lacked expertise in agriculture, policing, or "assisting newly elected provincial governments to prepare budgets and development plans, to obtain funding from the central government and to implement effective programs."[88] In contrast, civilian professionals possessed these skills but lacked the ability to safely access the provincial administrators.

By the provincial elections of 2009, the Iraqi government had benefitted from the reduction in violence brought on by the surge. However, by 2010, on the eve of US withdrawal, there was much the Iraqi government was ill-prepared to handle. The reintroduction of the Sunni populace into the political process challenged the power of the Shia and Kurdish politicians. In addition, despite exhortation by US officials and military officers, corruption remained rampant within the Iraqi government and Iraqi security forces. Moreover, another key shortfall of Iraqi governance was its inability to program, apportion, and allocate budgets. Lack of experience and clear authority, as well as corruption, siphoned off or froze up tens of millions of dollars necessary for reconstruction.[89] It was within this dysfunction that Iraqi and US negotiators agreed the US would complete its withdrawal of Iraq by 2011, unless a subsequent status of forces agreement (SOFA) could be decided.

Degrees of Embeddedness in Security Force Development
From 2003 to 2010, US forces relied on institution-influencing strategies to develop the Iraqi security forces. US forces employed advise-and-assist operations to develop security forces in Iraq. The result was an Iraqi security force that was only able to secure its juridical territory on its own for a little over three years after US withdrawal.

From 2003 to the start of the surge, the development of the Iraqi security forces proceeded through several evolutions governed by foundational prewar planning assumptions. Initially, it was assumed that the Iraqi security forces could be rapidly repaired. The repair and training of the Iraqi military was seen as a temporary and uncomplicated affair that could be handed off to contractors to allow US forces to withdraw rapidly.[90] After the Iraqi Army was disbanded by Coalition Provisional Authority Order #2 and the expectations of the Iraqi police failed to materialize, the training of Iraqi security forces transitioned from being a principally military, albeit largely ad hoc, affair initially to a more institutionalized effort.

In early 2003, Vinell, a defense contractor and the military training subsidiary of Northrop Grumman, correctly anticipated the need to repair and rebuild the Iraqi military after the invasion of Iraq. Vinell had previously helped train the Saudi Arabian army and won a $48 million, one-year contract in 2003 to train nine battalions of a thousand Iraqi soldiers each. Unlike Vinell's requirement in Iraq, its training of the Saudi

Arabian Army was not from tabula rasa conditions. Vinell's primary interactions with the Saudi Arabian military had been in training higher-level officers and facilitating operational-level planning and wargaming exercises.[91] They had no previous experience in creating a new institution altogether.

In August 2003, just a little over three months after the end of major conventional operations and about two months after Paul Bremmer disbanded the Iraqi military, Vinell began training the first Iraqi battalion. The training effort was subsumed under the hastily organized and deployed Coalition Military Assistance Training Team (CMATT). The nine battalions that Vinell was contracted to create in one year would constitute a light infantry division with twelve thousand soldiers, with more units to be trained later.[92] The first Vinell/CMATT trained battalions finished their initial training in October 2003. By April 2004, Operation Vigilant Resolve, the first battle of Fallujah, was developing. This was seen as an ideal opportunity to demonstrate the effectiveness of these newly minted Iraqi units by having them operate alongside US Marines. Indeed, the lack of sufficient US forces to clear and hold Fallujah intensified the request for employment of these freshly trained and equipped Iraqi units. However, the new Iraqi battalions refused to fight in Fallujah, even alongside US forces. By the end of April 2004, CMATT was disbanded.

The deactivation of CMATT and the establishment of the Multi-National Security Transition Command-Iraq (MNSTC-I) and NATO Training Mission-Iraq (NTM-I) under the same commander signaled a sea change in April 2004. The first commander of MNSTC-I and NTM-I was then Lieutenant General David Petraeus. In June 2004, the DoD's assessment of the state of Iraqi security forces was bleak. It determined that, despite nearly a year of CMATT training, no Iraqi units were able to plan, coordinate, execute, and logistically support their own operations or assume the lead for security in any area of Iraq.[93]

By September 2005, significant progress had been made under MNSTC-I and NTM-I as eleven Iraqi battalions were able to support Coalition forces during Operation Restoring Rights in Tal Afar. Iraqi forces outnumbered US forces for the first time in the war and were responsible for controlling their own battlespace. Operation Restoring Rights would be important for its ability to facilitate parliamentary elections later that year. However, its ultimate failure made it the ideal case study of COIN failure by 2007.[94] Operation Restoring Rights was held up as a success of the "clear, hold, build" methodology. However, when security elsewhere in Iraq demanded the redeployment of US forces, this left the trained Iraqi units to hold these freshly cleared spaces—and the city soon collapsed back into violence. As a result, Iraqi forces could only be expected to operate effectively alongside US forces. By 2006, they were still ill-prepared to hold ground, even ground previously cleared by US forces.

Throughout the time US forces advised the Iraqi Army, the chief weakness identified universally centered around logistics. This problem would not be ameliorated as late as the writing of this book, even with billions of dollars spent to develop the Iraqi logistical system. By October 2005, MNSTC-I and NTM-I reported that they were able to develop 115 battalions, 90 of which were evaluated as being able to plan, coordinate, execute, and logistically support their own operations without US support.[95] Even as late as 2014, this institutional weakness of logistics would again be noted by fleeing Iraqi soldiers. However, the logistical weakness of the Iraqi security forces was not and is not due to a lack of sufficient resources but a result of institutional corruption and systemic predation.[96]

Throughout 2005, the Iraqi Army continued to grow but was still unable to hold ground without US forces. On October 5, 2005, Lieutenant General Petraeus reported that 197,000 Iraqi security force personnel were trained, equipped, and ready to serve in support of the Constitutional referendum being executed that month.[97] Petraeus indicated that this equated to 115 battalions, with thirty-six of those at Level 2 (In the Lead) and the remainder at Level 3 (operating embedded in US units). In fact, he pointed out that seven of the trained battlions had their own areas of operation in Baghdad, and sixteen battalions were operating alongside US forces in Al Anbar. He also noted that Iraqi local police and the national police were also assuming greater responsibility for their areas. Despite all this success in late 2005, much of the terrain where the Iraqi security forces were in the lead would need to be re-cleared by US forces during the surge in 2006–2008.

In November 2005, there were more than 212,000 Iraqi security force members. By May 2007, the Iraqi security forces numbered almost 350,000.[98] Whereas the addition of over 100,000 more troops is significant, the numbers do not properly describe the degree of progress made. By July 2007, ninety-five Iraqi battalions were in the lead for security in more than half of Iraq's territory.[99] Though the US forces were still responsible for the bulk of the fighting in the most contested areas, Iraqi forces were only assigned to the least contested areas. However, by the time the bulk of the fighting ended in Iraq in 2009, the Iraqi forces were not primarily in the lead for any of this fighting. They were shielded from failure and used to hold the least challenging terrain. This did not serve them well three years after US withdrawal when the Iraqi Forces would have to fight ISIS on their own.

Advisor Surveys
From December 2016 to May 2017, twenty-two US advisors to the Iraqi security forces were surveyed as part of this book. These advisors served all over Iraq. Five advisors served before 2006, and seventeen served after 2006 but before the US withdrawal in 2011. The advisors are all commissioned officers serving in the US Army and Marine Corps. The lowest-ranking advisor surveyed was a first lieutenant, and the highest-ranking officer interviewed or surveyed was a four-star general. The advisors were recruited either because they possessed a foreign advisor military occupational specialty or because they were members of a military advisor association.

Advisors' Views

My survey of US advisors in Iraq produced an unexpected result. Despite the lackluster performance of the Iraqi Forces in 2014, most advisors who served in Iraq considered the Iraqi forces they trained to be tactically competent. Further, when advisors were asked who the better soldiers were or who had the better leadership—the Iraqi security forces or the insurgents—their responses were nearly a draw. Whereas few advisors took issue with the ability of the Iraqi security forces to operate tactically, a majority offered five common concerns which may help explain the subsequent failure of the Iraqi military in 2014: 1) systemic corruption, predation, and selfishness by senior officers; 2) logistical incapacity of the Iraqi military system; 3) near-universal inability to conduct lower and higher echelon maintenance; 4) lack of ability to understand and employ combined arms operations; 5) premature withdrawal of US forces.

Survey participants were first asked to evaluate the effectiveness of the Iraqi security forces they worked with at the time the advisor finished his tour (see figure 4). Of those surveyed, 72% said the Iraqi forces they worked with were "competent-with some exceptions," "generally competent," or possessed a "high level of competence."

Figure 4. Advisors' Perception of Iraqi Competence Upon Withdrawal.

These results are somewhat surprising given the reports of poor performance of the Iraqi forces facing ISIS in 2014. The potential for some degree of bias on the part of advisors assessing their own success exists. However, the failure of the Iraqi security forces in 2014 does not appear to have been largely tactical in nature. In fact, when Iraqi advisors were asked how easy or difficult it was to get Iraqi officers to adhere to basic tactical disciplines (such as patrolling, training, inspections, basic maintenance, advanced maintenance, post and relief of sentries, and operational security), the advisors gave the most positive assessment of Iraqi capacity in patrolling and training (see table 4). The results were similar with regard to getting the Iraqi units to conduct training.

Other tactical disciplines were not rated as highly. With regard to maintenance and inspections, 46% of the advisors reported that it was either "extremely difficult" or "impossible" to get Iraqi officers to do inspections; 46% said it was "extremely difficult" or "impossible" to get Iraqi officers to do basic maintenance; and 77% said it was either "extremely difficult" or "impossible" to get Iraqi officers to do advanced maintenance. Lastly, 63% said it was "extremely difficult" or "impossible" to get Iraqi officers to apply operational security (OPSEC) measures.

These are crucial assessments. First, the Iraqi security forces were trained and equipped to fight as a modern technological military force. In just one tank, there are literally tens of thousands of parts that may need to be repaired or replaced. Armored vehicles and technologically advanced weapons like anti-tank missile systems require significant hours of continuous labor to maintain. Failure to properly maintain weapon systems and vehicles, even those as simple as an M-16 service rifle or a HMMVV, can have catastrophic results in a modern military. Second, failure to protect the plans of operations and intentions for future operations increases the likelihood of and vulnerability to surprise, ambush, and destruction of a friendly force.

Table 4. Iraqi Officers and Tactical Disciplines.

	Impossible	Extremely difficult	Possible with effort	Possible- minimal effort	Easy-it came naturally
Patrolling	0%	14.29%	47.62%	28.57%	9.52%
Training	0%	27.27%	45.45%	22.73%	4.55%
Inspections	4.55%	40.91%	31.62%	13.64%	9.09%
Basic maint.	9.09%	36.36%	36.36%	13.64%	4.55%
Adv. maint.	31.82%	45.45%	18.18%	4.55%	0%
Post & relief	13.64%	18.18%	50%	13.64%	4.55%
Operational security	27.27%	36.36%	18.18%	13.64%	4.55%

Table 5. Iraqi Officers' Ability to Internalize US Warfighting Concepts.

	Impossible	Extremely difficult	Possible with effort	Possible-minimal effort	Easy-it came naturally
Mission Command	22.73%	27.27%	36.36%	13.64%	0%
Maneuver Warfare	19.05%	42.86%	23.81%	14.29%	0%
Combined Arms	0%	60%	20%	15%	5%
Fire and Maneuver	0%	25%	45%	25%	5%

Unlike tactical disciplines which deal with lower level tactical concerns, tactical concepts are more abstract philosophies or doctrinal warfighting concepts. Advisors were asked to rate how easy or difficult it was to get their Iraqi forces to internalize American martial concepts that form the foundation of US operational culture. These concepts include mission command,[100] maneuver warfare,[101] combined arms,[102] and fire and maneuver (see table 5). Though these more abstract concepts are difficult to assess precisely, it is instructive to note where advisors felt that teaching a particular concept was "extremely difficult" or "impossible." Of the advisors surveyed, 50% assessed that it was "extremely difficult" or "impossible" to get Iraqi officers to internalize mission command concepts that require an intense degree of trust between subordinate and commander and that delegate large amounts of authority. And over 60% thought the same of their ability to internalize maneuver warfare and combined arms concepts. The only concept that a significant majority of advisors estimated the Iraqi officers internalized well was fire and maneuver concepts with 80% saying it was "possible with effort," "possible with minimal effort," or "easy it came naturally."

The US style of war that was taught to the Iraqi military is based on maneuver warfare, which relies on centralized planning and decentralized execution. This decentralized execution depends heavily on the character and integrity of officers to build trust between superiors and subordinates throughout the entire organization. In order to develop the style of leadership necessary to execute this decentralized form of warfare, the leadership of officers and NCOs is developed and evaluated against sets of leadership traits and leadership principles. Among these traits and principles, advisors assessed Iraqi officers most positively with respect to judgment, courage, and endurance (see table 6). These qualities are crucial to effective leadership given the need for commanders to make rapid life-or-death decisions under duress.

Conversely, the two leadership traits the US advisors assessed Iraqi officers lowest in were integrity and unselfishness (see table 6). The qualities of integrity and unselfishness are essential for the creation of trust between subordinates and superiors. Unselfishness is a critical element in developing faith between a commander and subordinates. In the confusing and often terrifying context of combat, commanders who display the quality of unselfishness engender the willingness of subordinates to sacrifice. Where unselfishness engenders trust with one's subordinates, integrity is crucial for trust between an officer and his superiors. No superior could invest great faith and stock in the report of an officer who could not be trusted. This is particularly dangerous when the officer is claiming that he/she needs reinforcement, resupply, or permission to retreat. This quality of integrity is considered so vital to the US military that even as failures in physical and mental tests in entry level officer selection can often be remediated, lapses in integrity, even miniscule ones, most often result in immediate and permanent disenrollment from training.[103]

Given the poor performance of the Iraqi security forces against ISIS in 2014–2015, I had previously assumed that US advisors would evaluate the Iraqi insurgents as more capable than the Iraqi security forces. However, this was not born out by my research. When the advisors were asked who were better, the insurgents or the Iraqi forces, the results were a near draw. However, one particularly convincing point was made by an advisor regarding what happened when an Iraqi soldier defected to the insurgents:

> The soldiers/police were better soldiers because they possessed a semblance of formal training and command compared to the insurgents. Moreover, it is relatively common for a small percentage soldiers/police to leave their organization and join the insurgents. However, when they do so, they almost always assume a key leadership role in their local insurgent group. This serves as a key indicator that they are perceived as being better than a standard insurgent.

Table 6. Iraqi Officer Ability to Internalize US Leadership Traits and Principles.

	Impossible	Extremely difficult	Possible with effort	Possible-minimal effort	Easy-it came naturally
Justice	4.55%	22.27%	54.55%	13.64%	0%
Judgement	0%	18.18%	50%	31.82%	0%
Dependability	4.55%	18.18%	54.55%	22.73%	0%
Unselfishness	22.73%	31.82%	40.91%	4.55%	0%
Integrity	13.64%	36.36%	40.91%	9.09%	0%
Courage	0%	27.27%	27.27%	18.18%	27.27%
Endurance	9.09%	31.82%	27.27%	31.82%	0%
Know self and seek self-improvement.	18.18%	63.64%	9.09%	9.09%	0%
Be technically and tactically proficient.	4.55%	40.91%	40.91%	13.64%	0%
Set the example.	9.09%	45.45%	36.36%	9.09%	0%
Know your people, look out for their welfare.	9.09%	45.45%	31.82%	9.09%	4.55%
Seek responsibility, take responsibility.	22.73%	45.45%	22.73%	9.09%	0%

The Strategic Rentier State

This is important because if the Iraqi forces were better trained by the US and perceived as better even among the insurgents, then what accounts for their failure in 2014?

What is conclusive from my research is the near-universal assessment of the corruption and predation of Iraqi officers. The US advisors were asked about their perceptions regarding the possible corruption or predation of Iraqi officers targeted toward their subordinates, the government, or the Iraqi populace (see figures 5 and 6). Of those surveyed US officers, 55% assessed Iraqi officer conduct toward their subordinates and the government to be "more corrupt than what would be tolerated in the US military but not criminally so." And 63% described Iraqi officer conduct towards their subordinates as "more selfish than predatory." However, 36% saw Iraqi officer treatment of Iraqi troops as criminally corrupt by US standards, and 18% saw it as predatory. And 36% felt that Iraqi officer conduct with respect to the government rose to the level of criminal corruption by US standards. With regard to predation toward the government, 73% of US advisors surveyed saw the Iraqi officer behavior as more selfish than predatory, and only 9% saw it as predatory.

Follow-up questions were asked of the advisors asking them to describe the nature of any corruption or predation they witnessed. What is instructive about the advisors' responses is the difficulty that many had in translating their experiences with the Iraqi officers into their own cultural paradigm. Whereas significant numbers of those polled described Iraqi officer behavior as either "more corrupt than what would be tolerated in the US military but not criminally so" or "more selfish than predatory," almost all narrative responses provided would normally be considered as criminally corrupt in the US military.

Figure 5. US Advisor Perceptions of Iraqi Officer Corruption.

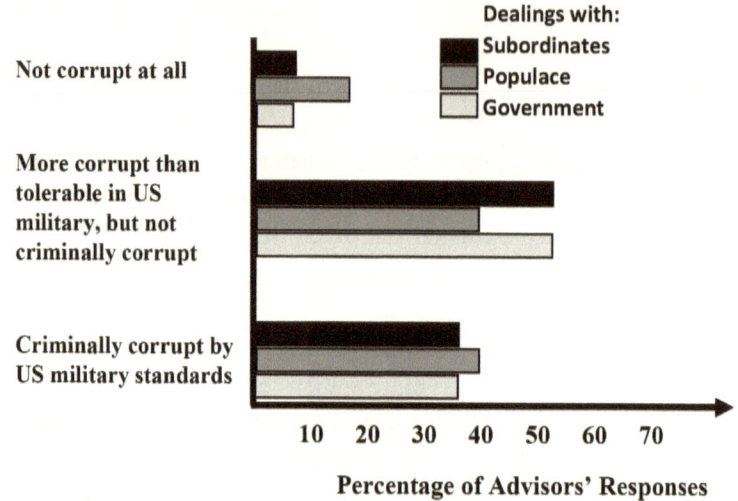

Figure 6. US Advisor Perceptions of Predation Among Iraqi Officers.

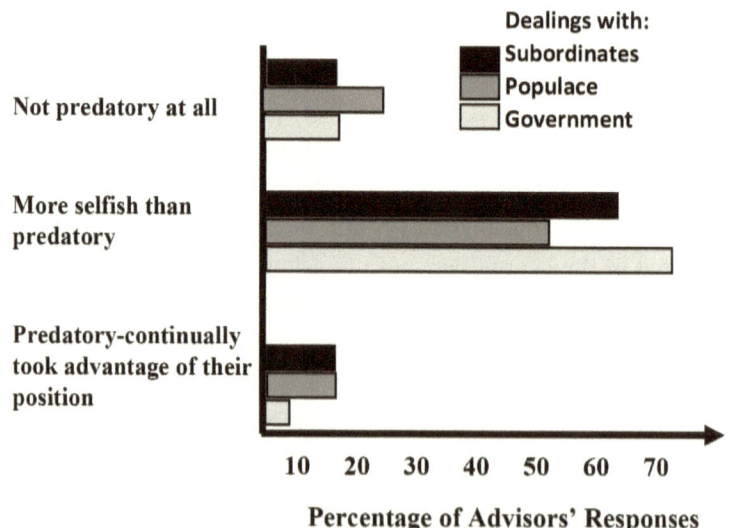

For instance, the most common anecdotal example of corruption and predation was the "skimming," "scalping," or "stealing" of subordinates' pay. In many cases though, this behavior was contextualized by the advisors citing that very often Iraqi logistics failed to reach the units, and the Iraqi officers had to pay for food, fuel, ammunition, weapons, and other such costs out of their own pockets. To pay for these necessities, officers often skimmed pay off from the soldiers' pay for the support of those same soldiers. Based on reports of the failed logistical system, this is almost certainly true. However, it is also illegal by US military standards, and almost just as certainly, it often had nothing to do with the care and feeding of troops.[104] Other experiences of corruption or predation included:

- treating younger soldiers as servants;
- "selling weapons, ammunition and supplies, sometimes to the enemy, for personal profit;"
- stealing food and supplies;
- failure to fire incompetent officers;
- receiving bribes and kickbacks;
- collecting pay for fictitious soldiers;
- forcing tolls on locals; and,
- "stealing, assault, manipulation, blackmail" and "torture and murder."

This critique of the differences between American and Iraqi cultures with regard to corruption is important. US advisors were called on to effect evolutionary change in the Iraqi Army in a revolutionary timeline. To do this, these advisors would have been required to materially change Iraqi culture within the military ranks in eight years' time.

Finally, the US advisors who participated in my survey were asked, "What one thing could have been done better by the US in Iraq?" The question was designed to be deliberately open-ended. It was instructive to observe how common certain responses were. Half felt the US had withdrawn precipitously before the evolutionary changes brought on by the US had a chance to take hold. The next three most common responses were represented by equal numbers of responses. The first was that the US had made a critical error in disbanding the Iraqi military in 2003. The second was that the Iraqi government had received insufficient US civilian support. Finally, an equal number of respondents argued that the US had instructed the Iraqi forces in a style of war for which the Iraqi military was unable to support with its own fire support and logistics. As a result, this made the Iraqi security forces permanently dependent on US support if they were to continue to fight as they had been trained.

First Test: ISIS 2014

At around 9:30 p.m. on July 22, 2013, guards at Taji Prison near Baghdad and Abu Ghraib Prison came under coordinated assaults. At Taji, the attack was stopped cold.[105] The raid in Abu Ghraib, however, facilitated the release of 500–1,000 senior Al Qaeda operatives.[106] In response, Prime Minister Maliki's interior ministry issued a statement that, "A preliminary investigation conducted by the crisis cell proved that there had been collusion between some of the correctional guards and terrorist gangs that attacked the prisons."[107] Whether proven or not, this claim was to be among the first of many indications that "some of the country's minority Sunni population feel increasingly marginalized by Prime Minister Nouri Maliki's Shia-led government" and were willing to work with ISIS to address it.[108]

On December 28, 2013, Maliki's government arrested another Sunni minister, Ahmad Al-Alwani, a member of the largest tribe in Al Anbar Province, the Dulaimi Tribal Confederation. The arrest took place at Alwani's home in Ramadi after nearly a year of Sunni protests against Maliki's regime. Many of these protests were led by the Dulaimi Tribal Confederation against what was seen as Shia repressive policies in Al Anbar Province. As a result of the arrest, some Sunni tribal militias began openly operating alongside Islamic State soldiers against Iraqi Army forces. By January 2014, ISIS forces and tribal militias had taken

Fallujah and much of Ramadi. In June 2014, ISIS officially declared their reconstitution of the Islamic caliphate under Abu Bakr Al Baghdadi as its religious and secular head.[109]

On June 4–10, 2014, perhaps no one was more surprised than the ISIS fighters themselves when they seized Mosul with only a handful of ISIS fighters against a garrison of two divisions, or nearly 30,000 Iraqi soldiers, not including Iraqi police. ISIS militants did not have to cordon the city and clear it with tens of thousands of fighters, tanks, artillery, and airpower. Rather, ISIS militants arrived at the city's outskirts in pick-up trucks and overran police checkpoints. ISIS militants then terrorized, mutilated, hanged, and beheaded police officers. Early in the fight, on June 5, 2014, Iraqi officers reportedly requested tank support. But when they refused to pay a sufficient bribe, they received an ancient Soviet-era tank with an untrained Iraqi crew instead of a US M1A1 main battle tank with US trained crew.[110]

As ISIS flags began going up on roof tops, panic struck the Iraqi forces. Frontline Iraqi accounts do not describe observing more than a few thousand ISIS fighters.[111] However, rumors of former Ba'athist officers helping ISIS sabotage the defense of Mosul were enough to cause Iraqi officers to abandon their posts en masse.[112] With the officers gone and Baghdad refusing to send material support, enlisted soldiers also fled; "Iraqi officials told the Guardian that two divisions of Iraqi soldiers—roughly 30,000 men—simply turned and ran in the face of the assault by an insurgent force of just 800 fighters."[113]

Surprised by their own success, ISIS forces exploited the weakness of the Iraqi security forces and began seizing more territory once Mosul fell. The day after Mosul's fall, ISIS forces entered Tikrit unopposed and ransacked the town. They also seized Baji, Iraq—home to Iraq's largest refinery. As more and more of Al Anbar fell to ISIS, the Baghdad government was incredulous and accused the Iraqi Army of betrayal. In an unpredictable turn, ISIS not only continued to fight, but they also began to govern. Though the ISIS style of government was particularly barbaric, "US officials have noted that ... the Islamic State's decision to hold and govern territory is a financial burden for the group, and thus a vulnerability that the United States could potentially exploit."[114]

By the end of 2010, Al Qaeda in Iraq had been defeated, and the Islamic State was at least marginalized. The US had created an Iraqi security force of approximately 350,000 troops with a large complement of modern military equipment and training. Yet, by mid-2014, with two divisions of Iraqi troops in Mosul, the city was overrun in days and held for years. What accounts for this?

CONCLUSION

From 2003 to 2010, the US relied on institution-influencing strategies, such as advise and assist, to develop Iraqi security and governance institutions. US advisors were required to effect evolutionary change in a revolutionary timeframe. They were required to do this through influence and encouragement but had no ability to compel. The result was to produce Iraqi security forces that were unable to act as the senior partner in an oligopoly of violence for more than three and a half years.

The US produced a strategic rentier state in Iraq. As long as Iraq continued to receive strategic rents in the form of close air support, advisors, and political supervision, it persevered. This was demonstrated in 2006–2010 and again since the US's re-intervention in 2014. However, once the US withheld strategic rents in 2010–2014, Iraq became what I term a crumbling state, characterized by low-to-no democracy and little-to no state-provided security in large swaths of the country until US forces re-intervened to stabilize Iraq in 2014. And should the US's provision of strategic rents be withheld again in the future, it seems likely that Iraq will revert to a crumbling state unless the Iraqi government changes or all threats to Iraqi governance dissipate.

The US developed the Iraqi government through advise-and-assist operations and transferred unilateral US political adaptations artificially in a top-down and outside-in fashion. The government that the US imposed on Iraq was far more liberal than even a more progressive state like the US could have tolerated at the time. The US instituted this alien government and transferred it in only the first fifteen months of a war that lasted eight years. The US transferred this government without any pretense of an oligopoly of the use of violence and without any meaningful access to local constituencies due to the security situation. The Coalition Provisional Authority rejected direct local elections, and later the Iraqi government also repudiated local governing councils.

When the Coalition Provisional Authority rushed the precipitous handover of the Iraqi government, this government lacked several crucial elements. First, the Iraqi Interim Government existed on paper only. It had no army or police at its disposal to enforce its laws. The Iraqi Interim Government had no capacity to extract taxes from its populace or to coerce citizens in the payment of these taxes. The Iraqi Interim Government had only the budget that had been given to it by the US. The Iraqi Interim Government also had no access to the citizens that it theoretically governed. And the representative nature of the Iraqi Interim Government was made even more illusory when nearly a quarter of the Iraqi population boycotted the political process.

The Iraqi Interim Government in many ways was not dissimilar from a government in exile—it existed behind, and rarely sallied forth beyond, the cloistered protection of the US Green Zone. Not only did the Iraqi Interim Government have no access to the populace, but it had no governance below the national level through which to govern locally. The US created a head of government that lacked any body. The US had created a *kafil* to provide the pretense of Iraqi sovereignty and legitimate its unilateral actions, instead of taking direct control transparently as it had during interventions before 1950.

Even after the successful elections of 2009, the Iraqi government only governed fairly and well until the US forces left. Almost immediately upon withdrawal of US forces, the Shia-led government began using state security organs to target Sunni political opponents. Little over three years later, the Shia majority government had disenfranchised significant portions of Sunni citizens in Al Anbar who began to align with ISIS and Al Qaeda.

The US military and the Iraqi insurgents were forced to adapt and negotiate in ways that the Iraqi government and security forces were not. In the face of potential defeat, US operations required continuous adaptation, exploration, and improvement that the Iraqi forces were not compelled to undertake. US forces also had to negotiate with the Shia and Sunni citizens directly as a result of a complex COIN strategy in ways that the Iraqi government was shielded from having to do.

Likewise, the Sunni insurgents had been compelled to continuously adapt as well. Even though the Sunni insurgents of Al Qaeda in Iraq and ISIS had been largely marginalized and defeated in many areas of Iraq by 2009, they were not completely eradicated, nor did they completely disappear. The Iraqi insurgents who would become ISIS had done battle with the world's most powerful military for years and were not completely destroyed. Therefore, whereas the insurgents were diminished in physical and material capacity by 2009, they had been forced to compete with, and in a sense benefitted from, what Michael Lewis of the Marine Corps University calls a training effect.[115]

In contrast, the Iraqi security forces and Iraqi government were shielded from these intense negotiations, competitions, and adaptations. Unlike the US forces and insurgents, the Iraqi security forces only took the lead in areas that had already been mostly secured by the US. Therefore, the Iraqi forces were largely shielded from the crucible of competition that existed between the US and the Iraqi insurgents by which both sides were forced to adapt and innovate. The pyrrhic US victory and unilateral adaptation, much of it technological and industrial, had a training effect on the Sunni insurgents who survived and would become ISIS; and it had a retarding effect on the Iraqi security forces and Iraqi government, which would become obvious upon US withdrawal.

The American style of war that US advisors artificially transferred to the Iraqi security forces requires a martial culture that depends on large amounts of trust, initiative, unity of effort, and individual capacity among officers. The US military has built this martial culture after decades of civilian oversight and strict enforcement of standards of conduct. Indeed, US military officers are still punished for infractions, such as adultery, that are not infractions in any other line of work in the US.[116] In contrast, US advisors were tasked with fundamentally changing an Iraqi military culture that had systematized what American military culture would describe as corruption and predatory behavior. US advisors were further tasked with effecting this change from outside the Iraqi military and strictly through advising and coercive influence. The advisors could only threaten to withhold their presence, fire support, logistics, and other such incentives to compel Iraqi forces. They could not, however, command and directly enforce standards. Finally, the US advisors were tasked with teaching a style of war that relies on unity of effort to a deeply divided and sectarian force.

The US style of war and its unique material and tactical adaptations made American adaptations difficult to sustain without perpetual US support. Many of the US adaptations were supported by massive infusions of technological innovations such as biometrics, UAVs, electronic warfare, armored vehicles, and persistent camera support. The style of US warfare taught to the Iraqis was heavily reliant on combined arms operations. The US's unique application of combined arms is heavily dependent on close air support which requires an industrial, technological, and tax base, which are well outside the capacity of the Iraqi government and most of the world's states. Unseen by most who observe the American war machine are also the tens of thousands of maintainers, trillions of dollars, and the colossal supply chain required to maintain the main battle tanks, armored vehicles, and plush command posts—all things the Iraqi security forces had grown accustomed to associating with the US war effort.

Iraq is a resource rentier state in the most basic understanding of the term. Iraq is also a strategic rentier state. Iraq stands to continue to receive strategic rents as long as the security situation deteriorates and as long as the US is willing to pay them. This creates a moral hazard where the Iraqi government only continues to receive these strategic rents so long as the security situation continues to deteriorate. The Iraqis actually lose these strategic rents if the security situation improves. This situation will perpetuate until one of two things happen—the US tires of paying these strategic rents in the form of combat air support, advisors, and financing and allows Iraq to implode as it did in Vietnam; or the Iraqi government and its security forces unilaterally decide to make fundamental governmental and societal changes.

Notes

1. Peter Maass, "The Toppling: How the Media Inflated a Minor Moment in a Long War," January 10, 2011, http://www.newyorker.com/magazine/2011/01/10/the-toppling; Bing West, *The March Up: Taking Baghdad with the United States Marines* (New York: Bantam, 2003).
2. "I Toppled Saddam's Statue, Now I Want Him Back," (July 5, 2016, http://www.bbc.com/news/world-36712233.
3. "Current Military Capabilities and Available Firepower for 2016 Detailed," (2016), http://www.globalfirepower.com/country-military-strength-detail.asp?country_id=iraq.
4. "Iraq Army Capitulates to ISIS Militants in Four Cities," (June 11, 2014), https://www.theguardian.com/world/2014/jun/11/mosul-isis-gunmen-middle-east-states; Daveed Gartenstein-Ross, "How Many Fighters Does the Islamic State Really Have" (February 9, 2015), https://warontherocks.com/2015/02/how-many-fighters-does-the-islamic-state-really-have/.
5. Estimates of ISIS fighters are around 1,000–1,500 and Iraqi Army (not including police) of approximately 30,000; see "Iraq Army Capitulates..." (2014) and Gartenstein-Ross (2015).
6. Patrick Cockburn, "Iraqi Government Recaptures Mosul Where it Suffered its Heaviest Defeat by ISIS," (July 9, 2017), http://www.independent.co.uk/news/world/middle-east/mosul-latest-iraqi-government-forces-recapture-city-heaviest-defeat-a7832186.html.
7. "Iraq by the Numbers," (December 19, 2011), https://www.dpc.senate.gov/docs/fs-112-1-36.pdf.
8. "Iraq: Corruption Perception Index," (2017), https://www.transparency.org/country/IRQ.
9. Nora Bensahel, Olga Oliker, Keith Crane, Richard R. Jr. Brennan, Heather S. Gregg, *After Saddam: Prewar Planning and the Occupation of Iraq* (Washington, DC: Rand, 2008).
10. Ibid., xix.
11. This is based on discussions between the author and Iraqi citizens in both Shia areas (Sadr City and Hillah in 2003) and Sunni areas (Ramadi 2008–2009) about what it was like to live under Saddam's security state.
12. Ibid.
13. United Nations Security Council Resolution (UNSCR) 1483 (New York: United Nations, 2003).

14. *Coalition Provisional Authority Number 2*. Baghdad, Iraq: Coalition Provisional Authority [CPA], 2003).
15. Ibid.
16. Bensahel, et al., *After Saddam*.
17. Ibid.
18. David A. Mueller, "Civil Order and Governance as Military Responsibilities," *Joint Force Quarterly* (2017): 43–50.
19. "Iraq's Most Wanted - Where Are They Now?" (September 1, 2010), http://www.bbc.com/news/world-middle-east-11155798.
20. *Measuring Stability and Security in Iraq* (2006), 24.
21. Patrick Cockburn, "Muqtada Al-Sadr and the Battle for the Future of Iraq," New York: Scribner, 2008), 202; Mitchell Prothero, "Baghdad Tense Amid Sadr Standoff," (April 6, 2004), https://www.upi.com/Defense-News/2004/04/06/Baghdad-tense-amid-Sadr-standoff/93301081268231/.
22. Bill Ardolino, *Fallujah Awakens: Marines, Sheikhs, and The Battle Against Al Qaeda* (Anaapolis, MD: Naval Institute Press, 2013); Daniel R. Green and William F. Mullen III, *Fallujah Redux*, (Annapolis, MD: Naval Institute Press, 2014); Richard Shultz, *The Marines Take Anbar* (Annapolis, MD: Naval Institute Press, 2013).
23. "Leave the wire" is a colloquialism employed by troops to denote leaving the security of larger US Forward Operating Bases (FOBs), or smaller Combat Outposts (COPs).
24. Predeployment training guidance delivered during training at Marine Corps Marine Air Ground Task Force Training Command (MCAGCC), 29 Palms, California, 2005.
25. The Operation was alternatively known to American Marines as Operation Phantom Fury.
26. John R. Ballard, *Fighting for Fallujah: A New Dawn for Iraq* (Westport, CT: Praeger, 2006), 58–78. Richard S. Lowry, *New Dawn: The Battles for Fallujah* (New York: Savas Beatie, 2010), 270–278; Thomas E. Ricks, "A Feaver-ish Take on the Surge in Iraq," (March 31, 2011), http://foreignpolicy.com/2011/03/31/a-feaver-ish-take-on-the-surge-in-iraq/, 400.
27. Bing West, *No True Glory: A Frontline Account of the Battle for Fallujah* (New York: Random/Bantam Books, 2005), 316; Ricks, "A Feaver-ish Take," 400.
28. Green and Mullen III (2014).
29. Jeffrey C. Isaac, Stephen Biddle, Stathis N. Kalyvas, Wendy Brown, Douglas A. Ollivant, "The New US Army/Marine Corps Counterinsurgency

Field Manual as Political Science and Political Praxis," *Perspectives on Politics* 6, no. 2 (June 2008): 347–360.
30. Peter Feaver, "Hillary Clinton and the Inconvenient Facts About the Rise of the Islamic State," (August 13, 2015), http://foreignpolicy.com/2015/08/13/clinton-surge-iraq-maliki-obama/; Ricks, "A Feaver-ish Take."
31. Isaac, Biddle, Kalyvas, Brown, Ollivant (2008).
32. Petraeus, "Counterinsurgency."
33. Lelah Khalili, "The Logistics of Counterinsurgency" (lecture, Georgetown University, Washington, DC, April 18, 2016).
34. FM 3-24, *Counterinsurgency* (2006), 2-2, uses "lines of operation" and this was the terminology used by US Marine Corps units in Iraq during the surge to describe unity of effort along multiple divergent lines of operation in COIN. However, lines of effort and lines of operation were used interchangeably in different areas of operation between the US Army and US Marine units. In the most recent *DoD Dictionary of Military and Associated Terms* (JP 1-02 2016), both terms are referenced.
35. Based on author's operational notes from 2008–2009.
36. Nagl, *Learning to Eat Soup With A Knife*, xiii.
37. Petraeus, "Counterinsurgency."
38. Karen DeYoung, "US Embassy Cites Progress in Iraq," (July 2, 2008), http://www.washingtonpost.com/wp-dyn/content/article/2008/07/01/AR2008070102860.html.
39. David Petraeus, *Report to Congress on the Situation in Iraq* (Washington, DC: Department of Defense, 2007).
40. Krepinevich, *The Army and Vietnam*, 236; Linn, *The Philippine War*, 325.
41. 461,700x.6=277,020, based on *Measuring Stability and Security in Iraq* (2006) and Amy Belasco, *Troop Levels in the Afghan and Iraq Wars, FY2001–FY2012: Cost and Other Potential Issues* (Washington, DC: Congressional Research Service, 2009), 12.
42. 33,400,000/277,020=120.569.
43. 33,400,000/287,040=116.36 based on *Measuring Stability and Security in Iraq* (2009) and Belasco, 12.
44. 33,400,000/141,100=236.711 based on Belasco, 12.
45. *Measuring Stability and Security in Iraq* (2006).
46. 33,400,000/461,700=72.341.
47. 461,700/ 168,754 square miles=2.7359.
48. McGrath, *Boots on the Ground*,193; Petraeus, *Report to Congress*.

49. 33,400,000/16,700=2000. Belasco, 12. General Petraeus noted that the troop surge got as high as 25,000 (e-mail communications, February 2019).
50. These forces refer to non-US and non-Iraqi members of the international Coalition fighting in Iraq but exclude security contractors.
51. 33,400,000/ 478,400=69.816. See note 49 of this chapter—if General Petraeus' estimate of 25,000 troops at the peak of the surge is used, the total force would be 486,700 and the resultant ratio would be 33,400,000/486,700=68.6.
52. Ghaith Abdul-Ahad, "From Here to Eternity," (June 8, 2005), https://www.theguardian.com/world/2005/jun/08/iraq-al-qaida; Huhanad Mohammed, "Iraq al Qaeda Militant Says Syria Trained Him," (August 30, 2009), http://www.reuters.com/article/oukwd-uk-iraq-syria-idAFTRE57T1OS2 0090830.
53. Mao, *On Guerilla Warfare*, 21.
54. Abdul-Ahad, "From Here to Eternity"; Mohammed, "Iraq al Qaeda Militant."
55. "US Foreign Assistance to Iraq," (2016) http://us-foreign-aid.insidegov.com/l/82/Iraq; "Iraq GDP 1960–2017," *World Bank* (2017), http://www.tradingeconomics.com/iraq/gdp: http://www.tradingeconomics.com/iraq/.
56. *US Overseas Loans and Grants: Obligations and Loan Authorizations, July 1, 1945–September 30, 2015* (Washington, DC: US Agency for International Development (USAID), 2015).
57. Luay Al-Khatteeb, "Iraq's Economic Reform for 2016," (December 13, 2015), https://www.brookings.edu/opinions/iraqs-economic-reform-for-2016/.
58. Anthony H. Cordesman, and Sami Khazai, *Iraq After US Withdrawal: US Policy and the Iraqi Search for Security and Stability* (Washington, DC: Center for Strategic and International Studies [CSIS], 2012); *Measuring Stability and Security in Iraq* (2006).
59. Loveday Morris and Missy Ryan, "After More Than $1.6 Billion in US Aid, Iraq's Army Still Struggles," (June 10, 2016), https://www.washingtonpost.com/world/middle_east/iraqs-army-is-still-a-mess-two-years-after-a-stunning-defeat/2016/06/09/0867f334-1868-11e6-971a-.
60. These survey results will be covered in detail later in this chapter.
61. Morris and Ryan, "After More Than $1.6 Billion."
62. Ibid.

63. Larry Diamond, "What Went Wrong in Iraq," *Foreign Affairs* 83, no. 5 (September–October 2004): 55.
64. Saad Jawad, *The Iraqi Constitution: Structural Flaws and Political Implications* (London: LSE Middle East Centre, 2013), 4–5.
65. Diamond, "What Went Wrong in Iraq," 39.
66. Ibid., 45–46.
67. Author's personal observations as a military representative to the North Ramadi, South Ramadi, and East Ramadi city councils from September 2008–March 2009.
68. Luciani and Beblawi, "Rentier State in the Arab World," 92.
69. Bensahel, et al, *After Saddam*, xxiii.
70. Jawad, *The Iraqi Constitution*, 8.
71. Bensahel, et al, *After Saddam*, xxiii.
72. Jawad, *The Iraqi Constitution*, 8.
73. Diamond, "What Went Wrong in Iraq," 53.
74. Bensahel, et al, *After Saddam*, xxiv.
75. "US Hands Over Power in Iraq," (June 28, 2004), https://www.theguardian.com/world/2004/jun/28/iraq.iraq1.
76. Ibid.
77. Ibid.
78. "Population Estimates and Voter Turnout for Iraq's 18 Provinces," (February 25, 2005), http://aceproject.org/ero-en/topics/electoral-participation/turnout/updatedelectionresults.pdf; "Iraq Election At-A-Glance," (January 24, 2005), http://news.bbc.co.uk/2/hi/middle_east/4033263.stm.
79. "Iraq Election At-A-Glance" (2005).
80. "Population Estimates and Voter Turnout for Iraq's 18 Provinces" (2005); "Iraq Election At-A-Glance" (2005).
81. *Measuring Stability and Security in Iraq* (2006).
82. Ibid., 11.
83. Ibid.
84. Brad Parks, "How a US prison camp helped create ISIS," (May 30, 2015), http://nypost.com/2015/05/30/how-the-us-created-the-camp-where-isis-was-born/.
85. "Provincial Reconstruction Teams in Iraq," United States Institute of Peace (USIP) (March 20, 2013), https://www.usip.org/publications/2013/03/provincial-reconstruction-teams-iraq.
86. Ibid.
87. *Measuring Stability and Security in Iraq* (2006), 10.

88. Ibid.
89. *Measuring Stability and Security in Iraq* (2006).
90. Steven Rosenfeld, "Iraq: Vinnell's Army on the Defensive," (January 21, 2004,), http://www.corpwatch.org/article.php?id=7842.
91. Ibid.
92. "US Firm to Rebuild Iraqi Army," (June 26, 2003), http://news.bbc.co.uk/2/hi/business/3021794.stm.
93. "The Iraqi Army: A Year of Progress," (June 29, 2015,), http://archive.defense.gov/home/dodupdate/iraq-update/Handovers/.
94. Based on author's personal experiences transitioning into the Iraqi theater. US forces coming into Iraq in 2008 were required to attend an in-country COIN academy. Tal Afar was used as a crucial case study to explain the risk associated with withdrawing US forces too early.
95. "The Iraqi Army: A Year of Progress," (2015), http://archive.defense.gov/home/dodupdate/iraq-update/Handovers/index.html; Bill Roggio, "Training the Iraqi Army—Revisited, Again," (October 7, 2005), http://www.longwarjournal.org/archives/2005/10/training_the_ir_1.php.
96. This will be expanded upon in detail later in this chapter as a result of 2017 US Advisor surveys conducted by the author.
97. Bill Roggio, "Training the Iraqi Army."
98. This includes local and national police. "The Iraqi Army: A Year of Progress," (June 29, 2015), http://archive.defense.gov/home/dodupdate/iraq-update/Handovers/.
99. Ibid.
100. *Marine Corps Doctrinal Publication 1—Warfighting* (MCDP 1) (Washington, DC: Headquarter Marine Corps, 1997), 87.
101. MCDP 1, 73.
102. MCDP 1, 94.
103. Based on author's personal experience serving as a student and instructor for Marine Officer Candidates School (OCS) and the Basic Officers Course.
104. Article 121 of the Uniform Code of Military Justice (UCMJ)-larceny and wrongful appropriation.
105. "Iraq Jailbreaks: Hundreds Escape in Taji and Abu Ghraib," (July 22, 2013), http://www.bbc.com/news/world-middle-east-23403564.
106. M. Abbas, "Al-Qaeda Militants Raid Iraq's Abu Ghraib, Taji Prisons," (July 25, 2013), http://www.al-monitor.com/pulse/originals/2013/07/iraq-al-qaeda-prison-raid-abu-ghraib.html.
107. Ibid.

108. "Iraq Jailbreaks."
109. John W. Rollins and Heidi M. Peters, *The Islamic State—Frequently Asked Questions: Threats, Global Implications, and U.S. Policy Responses* (Washington, DC: Congressional Research Service, 2015), 1.
110. Morris and Ryan, "After More Than $1.6 Billion."
111. "Iraq Jailbreaks"; "Iraq Army Capitulates to ISIS Militants."
112. John Beck, "Iraqi Soldiers Fleeing ISIS Claim They Were 'Abandoned' by Senior Officers," (June 15, 2014), https://news.vice.com/article/iraqi-soldiers-fleeing-isis-claim-they-were-abandoned-by-senior-officers.
113. "Iraq Army Capitulates to ISIS Militants."
114. Carla E. Humud, Robert Pirog, Liana Rosen, *Islamic State Financing and US Policy Approaches* (Washington, DC: Congressional Research Service, 2015), ii.
115. Lewis, "Replication and Extension."
116. Article 134 of the Uniform Code of Military Justice (UCMJ).

CHAPTER 4

THE TUMBLING STATE

NICARAGUA

> The US made the mistake in its original concept to create a supposedly nonpartisan constabulary in Nicaragua...Any attempt to create an honest, non-political military force without changing the nation's basic social and economic situation was probably impossible.
>
> –Richard Millett, *Guardians of the Dynasty*[1]

On January 2, 1933, in the port of Corinto, Nicaragua, US Marines boarded ships and brought Marine Corps leadership of the Guardia Nacional De Nicaragua (GN) to an end. The withdrawal ended five years of direct combat and twenty-one years of underwriting Nicaraguan security. The conflict resulted in 510 reported violent contacts, 75 Marines and Guardia killed in action, 122 Marines and Guardia wounded in action, an estimated 1,115 Sandinistas killed in action, 525 Sandinistas wounded, and another 75 Sandinistas captured.[2] The outcome though, according to Richard Millett, was that "the United States had given Nicaragua the best trained and equipped army it had ever known. But it also gave the nation an instrument potentially capable of crushing political opposition with greater efficiency than ever before."[3] What accounts for the creation of a Nicaraguan constabulary that effectively prevented major armed revolt for forty-six years?[4] What accounts for this success in a country that had averaged approximately one major armed revolt or battle every three years since independence? Further, what accounts for the failure of the US to achieve its goal of creating a nonpartisan Guardia Nacional? Finally, what accounts for the failure of the US to change the oligarchical and kleptocratic political culture at the root of Nicaragua's revolutions?

EXAMINING THE CASE—NICARAGUA

Seven years before the start of the US Civil War, the US would begin an eighty-year period of cyclical US interventions in Nicaragua. In February 1853, a detachment of US Marines and sailors from the *USS Cyane* landed for the first time in San Juan Del Norte, Nicaragua to "protect American lives and property."[5] Protection of American lives and property would be a common US Marine Corps and Navy mission during interventions in Nicaragua in 1853, 1854, 1857, 1867, 1894, 1896, 1898, 1899, and 1910 and would culminate in the maintenance of a Marine Corps Legation (embassy) Guard of 130 Marines for fourteen years from 1912 to 1926.[6] In 1821–1927, Nicaragua experienced over thirty-four armed revolutions or major battles over a period of ninety-three years. The only significant non-revolt periods occurred during the thirteen years of the Marine Corps Legation Guard. There was large-scale, political violence roughly every three years from 1821–1927. Indeed, in 1925, a supporter of the conservative party in Nicaragua told the American Minister to Nicaragua, Charles Eberhardt, that it was only by revolution that a party opposed to the government might hope to gain proper recognition in Nicaragua.[7]

From independence in 1821 until the 1990 elections, when the Sandinistas were largely voted out, power typically changed hands in Nicaragua in a violent and predictable pattern. The pattern generally consisted of a party, liberal or conservative, perpetuating itself in power with transitions of leadership internal to its party only. Simultaneously, the party out of power at the time would work vigorously and violently to remove the incumbent party from power. Often, the party out of power would lack sufficient force to unseat the incumbent party.

The incumbent regime would remain in power until riven by intraparty infighting when an incumbent executive refused to transfer power to other party members in their turn. The incumbent's unwillingness to share power even within his own party would instigate the rise of a military leader within the incumbent party. Without sufficient power to unseat the incumbent party leader on his own, the upstart military leader would often reach out to the opposition party and/or a foreign ally to help unseat the current dictator. As a result of this temporary alliance, the current dictator would then be replaced by another leader of the incumbent party. However, the process of unseating the dictator invariably exposed divisions within the incumbent party. Seizing upon this division, the opposition party would then rise up and violently unseat the disunited incumbent party. Then, the process would begin again almost immediately with the old incumbent party working to violently unseat the upstart.

This violent political process had repeated itself over and over again from 1821 until 1912. The US Marine Legation Guard was able to maintain stability for thirteen years from 1912 to 1925. At the end of those thirteen years, as Marines prepared to leave once again, the old process started anew once again.

On January 1, 1925, Carlos Solórzano and Juan Bautista Sacasa took office together as president and vice president respectively. They did so under the first combined conservative-and-liberal-party ticket. Upon the eve of US Marine departure in 1925, the US Department of State agreed to a postponement of the Marine withdrawal requested by the Nicaraguan government. This delay was conditioned on Nicaragua agreeing to the creation of a nonpartisan constabulary. In August 1925, the last Marines sailed away from Corinto, and within three weeks President Solorzano faced a mounting political crisis. General Chamorro compelled Solorzano to purge all liberal politicians in cabinet, and Vice President Sacasa fled the country.

From October 1925 to March 1926, Chamorro was the de facto executive of Nicaragua, even if Solorzano was de jure the president. However, on March 13, 1926, President Solorzano resigned, and Chamorro formally replaced him. From May to September 1926, the liberals pounced on the weakness caused by the conservative party's infighting and revolted. By September 1926, the Nicaraguan government was bankrupt, and Chamorro was forced to resign. In November 1926, Chamorro was replaced by a fellow conservative, Adolfo Diaz. Diaz was supported by the US almost exclusively because his opponent, the liberal Sacasa, was supported by Mexico. From November 1926 to May 1927, the liberals fought an effective campaign against the Diaz regime and his forces. However, the US protected the Diaz regime from complete failure, and by the spring of 1927, Nicaragua was on the verge of anarchy and starvation. It was under these conditions that the US intervened once again in Nicaragua.

Almost a year to the date after the advent of the Liberal uprising in 1926, all the Liberal Generals had signed the pact known both as the Stimpson Peace Plan and the Pact of Espino Negro. After months of negotiating from September 1926 until the US minister Henry L. Stimson's arrival in April 1927, Nicaragua had slipped deeper into anarchy. Nicaragua risked national famine if the wheat harvest was not brought in by June. By May 1927 the Nicaraguan government was bankrupt, and Chamorro was forced to resign. On May 12, 1927, all the liberal generals—except Augusto Cesar Sandino—signed the Pact of Espino Negro under the conditions of a US-supervised election that the liberals would ultimately win. Under the terms of this pact, the liberals and conservatives agreed to three key terms: First, all non-official state security forces would disarm; second, all parties would accept the results of elections to be held in 1928; and third, all parties would support the creation of a nonpartisan military force under US supervision.[8]

US policymakers hypothesized that nonpartisan security forces were a cause of good governance and not an effect of it.[9] After intervening numerous times in 1821–1927, US policymakers mistakenly identified Nicaragua's security forces as the source of Nicaragua's perpetual armed revolts. Dr. Dana Munro, the charge d'affaires in Nicaragua, argued that, "The old armies were or seemed to be one of the principal causes of disorder and financial disorganization. They consumed most of the government's revenue, chiefly in graft, and they gave nothing but disorder and oppression in return. We thought that a disciplined force, trained by Americans, would do away with the petty local oppression that was responsible for much of the disorder."[10] What was ignored was the predatory political system that coopted the Nicaraguan security forces to alternately unseat or install one liberal and conservative dictator after another.

ENEMY-CENTRIC COIN—WAR OF MANEUVER

Enemy-centric COIN methods do not adequately explain the success of the US Marine Corps in creating the Guardia Nacional. Nor do enemy-centric methods explain the Guardia's ability to secure Nicaragua for forty-six years after US withdrawal. Enemy-centric COIN methods were present and weighted during the US Marine intervention, but they were not effective. The Sandinistas were every bit as capable a fighting force when the Marines left in 1933 as they were when the Marines landed in 1927. Whereas the Sandinistas could not compete effectively with the Marines and the Guardia Nacional in conventional war, in guerilla warfare the Sandinistas demonstrated substantial tactical acumen. The Sandinistas also successfully protracted the campaign. In addition, even though the Sandinistas were never able to extend their influence beyond their own territory, the Marines and Guardia were never able to completely eliminate them as a battlefield threat. The Sandinistas were "successful" inasmuch as they were not destroyed. The Marines were successful, not in their enemy-centric COIN methods but in their ability to create a security force that could secure Nicaragua for over four decades after the Marines left.

Nicaraguan security forces had been so diminished by late 1927 that they could not protect the elections or provide even the most basic security responsibilities. As a result, in 1928, the Marine Corps landed a series of ground combat, artillery, and aviation units under the 2nd Marine Brigade. This represented the largest deployment of Marines since the end of World War I.[11]

From the onset of the conflict, Sandino's forces demonstrated a complete inability to compete against the US Marines in conventional warfare. In the very first fight against the Marines and Guardia, Sandino's forces greatly outnumbered the three Marine officers and fifty new Guardia. Nevertheless, the Sandinistas were soundly beaten by the tiny Marine and Guardia forces.[12] Furthermore, during contacts on September 19, October 9, October 26–27, November 1, and December 5, 1927 and January 1, 1928, the Sandinistas lost 260 fighters killed in action.[13] And, in every case, the Sandinistas far outnumbered their Marine and Guardia Nacional foes. In contrast, the Marines and Guardia only lost eight Marines and six Guardia killed in action.[14] Though the Sandinista losses are estimates and may suffer from exaggeration, they nevertheless correspond well to Sandino's decision to move away from costly conventional warfare and toward guerilla warfare.

Whereas the Sandinistas could not compete effectively in conventional warfare against the Marines, they did, however, demonstrate tactical savvy in guerilla warfare. After his conventional losses, Sandino adopted the tactics of protraction and guerilla warfare to "win" in the long term by simply remaining relevant and not being destroyed. Therefore, during the Marine and Guardia offensives in support of the 1928 elections, Sandino avoided their main forces and struck in undefended regions. This did little to defeat the Guardia strategically, but the constant raids maintained Sandino as a relevant threat.

In line with the tenets of General Order 100 from the US Civil War and the Philippine-American War, Marines and Guardia Nacional de Nicaragua declared martial law. They targeted those actions "approximating organized warfare" as well as normal criminal activity.[15] Martial law gave the Marines and Guardia authority to press local guides into service against their will. It also allowed the Marines and Guardia to arrest or detain not only insurgents but also those who provided material support. Finally, the Marines and Guardia were authorized to confiscate or destroy such personal property and take any measures that will "injure the cause of the bandits and lead to their defeat" or cause insurgents to forsake their "banditry."[16] Nevertheless, even as these enemy-centric COIN methods may have pressured the Sandinistas, they did not materially diminish their combat effectiveness.

POPULATION-CENTRIC COIN

Population-centric COIN methods were present, but they were not weighted nor can they really be considered effective by the Marines in Nicaragua. This is unusual, because—despite the lack of substantial population-centric COIN efforts—Nicaragua nevertheless survived almost five decades after the US Marines withdrew in 1933. Population-centric methods were not as critical in the US intervention in Nicaragua. Population-centric COIN calls for the co-option of the populace by the counterinsurgent force away from the insurgents. However, revolt in Nicaragua, at least until well into the reign of Anastasio "Tachito" Somoza DeBayle in the 1970s, was almost exclusively a violent struggle between small kleptocratic political elites—not popular uprisings. These near-feudal elites had small loyal followings and primarily forcibly impressed peasant soldiers into their ranks. According to Thomas Walker and Christine Wade, there was no real mass mobilization of the peasant population that the US had to contend with.[17] Rather, the US Marines and Guardia Nacional worked primarily to eliminate popular support for Sandino and his forces in Sandinista territory. This effort was not successful.

Walker and Wade note that "throughout Nicaraguan history, a small elite controlled most of the means of production and garnered most of the benefits. The country's rulers—whether openly dictatorial or ostensibly democratic—almost always governed on behalf of the privileged few."[18] Walker and Wade's analysis of the Nicaraguan populace is valuable because it gives some idea as to how such relatively small governmental and revolutionary forces, never more than 5,000 on either side, could easily dictate the fortunes of almost 700,000 Nicaraguans.[19] Walker and Wade point out that, unlike other Latin American states, Nicaragua is not overpopulated and it has "no major racial, ethnic, linguistic, or religious divisions."[20] Moreover, they argue that as the majority of the populace are Spanish Mestizos,[21] there is little racial prejudice, and nearly all Nicaraguans are Catholic. Therefore, with the exception of political ideology and wealth inequalities, there appear to be few of the typical causational factors of instability[22] that are present in many intrastate wars. Yet, as previously demonstrated, Nicaragua has been continuously roiled by armed internal revolt.

This understanding of who constituted Sandino's supporters, or the Army for the Defense of Nicaraguan Sovereignty (Ejército Defensor de la Soberanía de Nicaragua-EDSN), is essential to analyzing the US intervention from a COIN perspective.[23] The bulk of Sandino's support was anchored disproportionately in Nueva Segovia. This support was precipitated in the same manner many insurgent movements gain support. Some of Sandino's supporters were genuine volunteers supporting their padrone. They initially supported him against the conservatives in 1927 and later against their own liberals and the Americans in 1928–1933. The rest of Sandino's troops were feudal conscripts who were impressed into service by Sandino, their feudal lord, just as their forbears had been Shanghaied into previous revolts in 1821–1933.[24]

Sandino and his followers had little national support. With the exception of the feudal popular support of the populace of Nuevo Segovia, Sandino was not able to extend his influence beyond this region. As Millett notes, "five years of fighting the Marines failed to eliminate Sandino, but, Sandino, except for an occasional daring raid, had largely been confined to isolated and thinly populated sections and had never taken a major city."[25]

However, because Sandino enjoyed strictly local support, the Marines pursued a regional population-centric COIN campaign. With the possible exception of Chesty Puller's Company M of the Guardia, efforts to co-opt the populace of Nuevo Segovia away from the Sandinistas were not successful. US Marines were very familiar with insurgency and counterinsurgency operations as a result of service in the Philippines, Cuba, the Dominican Republic, Haiti, and Panama. As such, Area Order-3, 1931, issued by the Chief of Staff for the Guardia Nacional de Nicaragua, Colonel Julian C. Smith, demonstrates a clear understanding of population-centric COIN. Even in areas that were under martial law, Smith admonished Marines and Guardia that,

> Every effort will be made to assure them of the friendliness of the Guardia toward all peaceable citizens and to extend such protection as will enable them to carry on peaceful occupations without molestation and to accumulate property without fear of robbery.[26]

Notwithstanding this, in practice, the Marines and Guardia Nacional failed in executing some population-centric methods. They attempted a resettlement program in August 1929 to isolate the Sandinistas from the populace. However, the attempt was poorly planned and executed and actually backfired. Due to the poorly planned and executed nature of this plan, rather than drying up support for the Sandinistas, it actually rejuvenated the then faltering Sandinista movement in Nuevo Segovia.[27]

Troop Ratios

Guardia and Marine counterinsurgent-to-populace ratios never got within even the most liberal estimates of forces required. Contemporary estimates of required counterinsurgent-to-population ratios range between 1:50 as the most conservative and 1:91 as the most liberal. At the height of the US intervention, there were approximately 5,000 Marines and sailors associated with the 2nd Marine Brigade. There were 1,710 troops associated with the Guardia at this same time.[28] This provided a force of 6,710 total troops to secure the 1928 elections. With a population of around 670,000[29] this gives a ratio of 1:100 counterinsurgents to population. By 1931, the ratio dropped to 1:233 and then to 1:295 counterinsurgents to population by the end of 1932. If Krepinevich and Linn's idea of foxhole strength are applied, then the force ratios may be further lowered by as much as 40 percent more to 1:140 at the height and 1:326 by 1931 and 1:413 by 1932.[30] As such, counterinsurgent-to-population ratios were well outside what are considered normal tolerances for success. Yet, the US-supported state nevertheless endured long after US withdrawal.

Counterinsurgent-to-terrain and counterinsurgent-to-insurgent ratios also do little to explain the long-term success of the Guardia after US withdrawal. At the peak troop levels, this provided 7.5 Marines or Guardia per square mile.[31] This low troop-to-terrain ratio was further exacerbated by the existence of insurgent sanctuaries across the border in Honduras and Costa Rica. Further, the Sandinistas ranged in estimated size from a low of approximately 500 to an estimated high of 5,000.[32] Based on the numbers of casualties, enemy sightings via ground and aerial reconnaissance, the number of weapons turned in, and the forces described by Sandino himself, there is reason to assume that the Sandinistas ranged between 3,000–4,000 full and/or part time troops. Even if the lower of these is assumed at 3,000, this provides a near parity with Marine and Guardia forces. In 1928, the Guardia and Marines would have had a 2:1 advantage.[33] However, that advantage would have been diminished by 1930 where there were 1,870 GN and 1,500 Marines (3,370 total) giving the Guardia and Marines only a 1.3:1 advantage. Then, assuming the insurgent numbers held steady or increased, there were 2,370 Guardia and 500 Marines (2,870 total) in 1931 which gave the insurgents near parity or a slight advantage. Then, in 1932, with the end of the intervention and only 2,274 Guardia remaining behind, this provided the insurgents with a solid advantage. This is supported by Charles Neimeyer who asserts that "either due to arrogance or a dearth of Marines able to take the field, Marine forces sent against Sandino and his men were always understrength and outnumbered."[34] This fact is further borne out by the detailed accounting of all 510 contacts between the Marines and Guardia and the Sandinistas in the official US Marine Corps archives.[35]

Enduring Insurgent Sanctuaries

At no point during the intervention did US Marines or Guardia permanently foreclose Sandinista sanctuaries. Sandinista sanctuaries allowed the insurgents to survive and slowly bleed the Marines and Guardia. However, these sanctuaries also prevented the Sandinistas from succeeding. Sandino's forces were able to stay alive and remain relevant principally because of unsecured borders with Honduras and Costa Rica. Still, even though these sanctuaries allowed the Sandinistas to remain alive, they also kept them isolated from the population they meant to control and influence.[36] As a result, the Sandinistas had little direct daily impact on the 1928 and 1932 elections or the day to day functioning of the state.

Continued Support After Withdrawal

Enduring combat support and military economic aid were not essential in the survival of the Somoza regimes after US withdrawal of combat formations. Instead, economic and military support was used to guarantee Nicaraguan support of American foreign policy. The Somozas shrewdly aligned themselves against the same enemies as their patron, the US. In World War II, Nicaragua declared war on Germany, Italy, and Japan just days after the US did. As a result, Nicaragua benefitted from millions in Lend-Lease Agreement dollars. During the Cold War, the Somoza regime became virulently anti-communist and even offered to send troops to Vietnam in support of the US war effort there. Consequently, Nicaragua received military equipment and training. The support provided to Nicaragua was not to prevent imminent Nicaraguan state failure, but rather to prevent interference by the Axis Powers during World War II and communist powers during the Cold War.

Governance and Security Force Development

Degrees of Embeddedness in Governance Development

Nicaragua provides a unique case study among the four selected for this book. Nicaragua represents the one case in which different degrees of embeddedness were combined to develop Nicaraguan governance and security. The divergent degrees of embeddedness that were employed provide insight not only into intervention outcomes with respect to SLAW but also outcomes in host-nation governance and security.

The US deeply embedded its officers within and inhabited the Guardia Nacional relying on institution-inhabiting strategies. In contrast, the US relied exclusively on influence alone to develop or remediate Nicaraguan governance. In point of fact, the US had no formal program at all to ameliorate Nicaraguan governance. Instead, US policymakers sought to deflect imperialist critiques that it had appropriated Nicaraguan sovereignty. The efforts to maintain at least the pretense of Nicaraguan sovereignty resulted in a sovereignty fallacy.

From 1850 until the present, the US Department of State has preferred to use coercive influencing relationships to advise and assist the Nicaraguan government in its development. Whereas the US assumed full control of the governance of other states, such as the Philippines, Cuba, Panama, Guam, and Puerto Rico, the US shied away from this policy in Nicaragua. In Nicaragua, the US eschewed full political control and publicly proclaimed its support for the total sovereignty of Nicaragua. However, this pretense of sovereignty, or sovereignty fallacy, flies in the face of the unmitigated veto power that the US possessed in Nicaragua.

The first example of this dissonance between US control and the pretense of Nicaraguan sovereignty can be discerned even before the

Stimson Peace Plan was finally approved by the liberals. On May 8, 1927, as plans were being made to identify who would lead the Guardia Nacional de Nicaragua's training effort, US agents preferred to rely on US Marines. This was due to the Marines' history and familiarity with Nicaragua. It was also due to the preference of the US Navy officers who were in the lead for Nicaraguan operations. However, President Adolfo Diaz preferred to use US Army personnel. Not only would Nicaraguan officials have no say over which US military service trained the Guardia, they would also have no say over which Marine officers were selected for the duty. Identifying the American units and officers involved in training would be wholly decided by US officers. Therefore, in the end, President Diaz was overruled, the US Marines were chosen, and the individual officers were selected by that service.[37]

A second example of dissonance was that the US Marines needed no specific Nicaraguan legal mandate to initiate and maintain lethal operations in Nicaragua. The Marines reported their first lethal contact with the Sandinistas on July 16, 1927, in the Ocotal region of Nicaragua. Even though Nicaraguan politicians had verbally requested that the US forces intervene, it was not until June 18, 1927, that the Marines received their first formal written guidance from President Diaz. It was only on July 30 that they finally received a formal decree, and it was over a year and a half before Munro-Cuadra Pasos Agreement was ratified on December 22, 1928.[38] This clearly demonstrated that host-nation, sovereign legal mandates were superfluous to US lethal operations.

A third example of the disconnect between the pretense and the reality of Nicaraguan sovereignty lay in the power to establish command relationships. The prerogative of establishing command relationships between military forces rested with the American military and not the Nicaraguan state. Command relationships are foundational to any military operation and a fundamental prerogative of state sovereignty. In mid-1927, the Nicaraguan Army was being disbanded. The municipal police were subsumed into the Guardia Nacional, and the existing Guardia were being led by Marines. Operating alongside these mostly indigenous forces was the 2nd Marine Brigade. With such disparate forces operating all across Nicaragua against an ill-defined insurgent foe, there was a consequential need for clearly defined command relationships between these forces. Yet, it was the commander of the US Navy Special Service Squadron—not the Nicaraguan government—who established the command relationships for all these elements involved in combat and policing duties.[39]

A fourth example of the disconnect between the pretense and reality of Nicaraguan sovereignty lies in the budgetary control of Nicaraguan security forces. The paramount control that a state demonstrates over its security forces is allocating the budget of these forces. From 1927 to1933, apart from being required to fund the Guardia Nacional's budget, the Nicaraguan government had little involvement or say in it. American agents sought to keep the budgeting and disbursement powers out of Nicaraguan hands for fear of systemic graft in the Nicaraguan system. Key areas for graft were in the allocation of contracts, the disbursement of pay, the purchase of quartermaster goods, and the like. American commanders of the Guardia estimated the required budget of the Guardia and submitted this to the Nicaraguan president. The president could accept this budget or negotiate a smaller budget. However, with the exception of negotiating the size of the budget and providing the money for it, the Nicaraguan government had little to do with the Guardia Nacional's budgetary affairs.[40]

A fifth example of a sovereignty fallacy at work is the Nicaraguan elections. The Nicaraguan government under President Diaz was asked by the US State Department to cede complete control of Nicaraguan elections over to US Army General Frank McCoy. Both Sandino and his conservative foe, President Diaz, possessed legitimate concerns about the fairness of any election. As a result, General McCoy was given complete control over the election machinery, even to deciding whether Marines or Guardia personnel would be inside the polling sites. The Nicaraguan government had little say in key decisions regarding the election. As Dana

Munro writes of his experience during this period, "the chairmen of the 13 departmental electoral boards were for the most part officers from the United States Army".[41] The elections were a success by every measure. Even the elections' losers claimed it was "the cleanest in the country's history".[42] The turnout for the election was almost 90 percent among registered voters.[43] However, the election was planned and executed completely independent of Nicaraguan government control.

Degrees of Embeddedness in Security Development
The first American effort to create a nonpartisan Nicaraguan security force began with the Managua Metropolitan Police Force. The US Marines' landing in 1910 was the ninth such landing in fifty-six years. Marines would thereafter be stationed at the American Legation in Managua in 1912–1926 to forestall the need for another such landing. However, early on, the question arose as to how the Legation Marines could eventually withdraw without precipitating a return to the previous cycle of violence.

To effect this change, the US Department of State demanded a US agent be assigned to the Managua Metropolitan Police Force. The Nicaraguan police were largely seen as strictly instruments of political repression designed not to enforce law but to repress political opposition. Greg Scull, a former member of Theodore Roosevelt's Rough Riders, was selected for the task of professionalizing this force. Scull assumed the role of Inspector General and Instructor on February 5, 1912. In less than ninety days, he resigned in embarrassment and frustration.[44]

As a result of the long presence of the Legation Guard and the failure of the Scull affair, Nicaraguan security forces not only stagnated but they nearly disintegrated altogether in 1912–1925. A report from the Marine commander of the Legation Guard to the Commandant of the Marine Corps, dated August 26, 1924, related that Nicaragua had the smallest military budget of any Central American state at that time. On its face, this fact seems indicative of the relative stability achieved during the tenure of the Legation Marines. However, Neimeyer, Gravatt, and Millett report that the stability produced during this period was artificial and temporary.[45] They argue that because the Legation Marines were the real source of security, once they were withdrawn the Nicaraguan police and army were incapable of securing Nicaragua on their own.

In 1925, the Nicaraguan government contracted Calvin Carter, a US Army major, to act as Jefe Director of a Guardia Nacional. In spite of significant obstacles, Major Carter and his four American assistants made significant progress in developing this original Guardia from July to October 1925. His initial obstacles came as Nicaraguan politicians demanded that the Guardia Nacional be maintained under the Ministry of War and the quartermaster department. Major Carter and the US Department of State vigorously resisted this. They were concerned by the considerable opportunity for graft if Nicaraguan officers were given control of US logistics provided to the Guardia.

Losing this battle, Carter went on to demand that recruits be volunteers, professionally trained, and provided uniforms, regular pay, and subsistence. However, this would all prove difficult. The middle and upper classes in Nicaraguan society had little interest in enlisting, and the army was poorly regarded among the lower classes.[46] Therefore, of Carter's initial pool of 200 recruits, almost half could not pass the baseline physical examination. Making matters worse, the Guardia Nacional had an inadequate budget and no machine guns. Still, Carter made solid progress until the start of the Chamorro-Rivas Revolt in October 1925.

Carter's constabularies had done an admirable job of resisting partisan subversion. Nevertheless, the Guardia Nacional was simply not ready to play a decisive role in preventing armed revolt. By the time the Legation Marines withdrew in August 1925, Carter's Guardia Nacional had increased in size to eighteen officer cadets and 225 enlisted men. Carter had originally assumed he would have two years to develop this force. However, his nascent Guardia Nacional was thrown into battle less than three months after being formed.

General Chamorro had been intensely antagonistic toward the Guardia before coming to power—when the Guardia had prevented him from taking control. Before coming to power, Chamorro had previously and aggressively tried to defund the Guardia Nacional. Once in power though, Chamorro began strengthening the Guardia with the same fervor with which he had formerly tried to weaken it. Millett argues that the creation of Carter's Guardia Nacional "signaled a turning point in Nicaraguan military history. Nonpartisan it was not, but it was better disciplined, better trained, and better led than any previous Nicaraguan force."[47] During the May uprising, the Guardia Nacional was sent into battle alongside the Nicaraguan Army. The Army was twenty-five times the size of the Guardia Nacional, but the Guardia was employed to such a degree that it suffered casualties at a rate of fifty-two times that of the Army. This, for all intents and purposes, was the end of the Guardia Nacional until the arrival of the Marines again in 1927.

From the re-intervention of US Marines in 1927 to the withdrawal of the Marines in 1933, the growth of the Guardia Nacional was impressive in every respect except one. The Marines initially conducted the preponderance of combat operations against Sandino until the end of the 1928 elections. From that point on and until 1933, the Guardia Nacional would gradually assume responsibility for most operations. Within only a year of establishment, the Guardia Nacional would never again have difficulty recruiting and training enlisted men. The same could not be said about the US Marines' production of Nicaraguan officers.

The Marines were able to grow the Guardia Nacional 500% from disbanding the army in 1927 to a full complement of enlisted Guardia by December 1932.[48] This success was in spite of near-continuous lethal contact with a capable foe and perpetual budget battles. The Guardia Nacional had needed to grow quickly and professionally and under a number of extenuating circumstances. One of the more challenging circumstances was the fact that the majority of the early Guardia Nacional recruits in 1927 could not pass basic physical exams or literacy tests. The matter became so problematic that American Marines had to establish Spanish schools to help teach Spanish speakers how to speak, read, and write Spanish. To further compound the matter of recruiting and training, syphilis was rampant among recruits and Guardia enlisted. In spite of these issues, the Guardia Nacional grew and became increasingly more potent and effective.

To understand how far the Guardia Nacional improved in 1927–1933, it is important to understand the original state of affairs in the Nicaraguan security forces. First, leadership in the Nicaraguan military was feudal in nature before the Marines began developing the Guardia Nacional; that is, Nicaraguan officers were given ranks according to the size of forces they could raise or impress from their home regions and the degree of their connectedness to the current regime.[49] This is not to say that gifted leaders did not exist. However, their promotion was not based on merit per se. Indeed, because an officer received higher rank based on the size of force he could recruit or impress and pay, this also incentivized graft as a means of paying soldiers.

Additionally, the sheer number of generals in the relatively tiny Nicaraguan military pre-Marine Corps intervention in 1912 is suspect. For instance, in 1908, there were seventy-seven Nicaraguan Army generals to command a force of no more than 5,000 soldiers.[50] To provide some perspective, the current size of the United States Marine Corps—the smallest in the Department of Defense—is 185,000 Marines, and it is only statutorily allowed 60 flag grade officers, or one general officer per 3,083 Marines. The Nicaraguan Army had had one general officer per sixty-five soldiers—or roughly a general per large platoon.

Next, members of the Nicaraguan security forces were usually not volunteers and professionals before the Marine intervention. They were generally conscripted and indentured serfs and often serving against their will. Richard Millett records a report from the Marine Corps Historical Association that, "While the law provided for universal military service, in reality the ranks were filled by forcibly rounding up members of the lower class, tying them together, and shipping them off to the nearest army camp."[51] Nicaraguan recruits were conscripted from the barefoot classes, and only the poorest citizens were pressed into service. Nicaraguan soldiers generally had to subsist by looting during war. Even when they received regular pay from the government, their officers withheld portions of this ostensibly for the purchase of their rations but certainly for self-enrichment as well.[52]

When the Marines landed in 1927, Lieutenant Colonel Elias Beadle (USMC) was promoted to the rank of Major General and Jefe Director of the Guardia Nacional de Nicaragua. US Marines were installed as officers within the Guardia Nacional to train and lead the Guardia in combat. By October, the Guardia Nacional had grown to 438 Guardia personnel (see figure 7).[53]

Figure 7. Comparative Guardia Growth, 1927–1933.

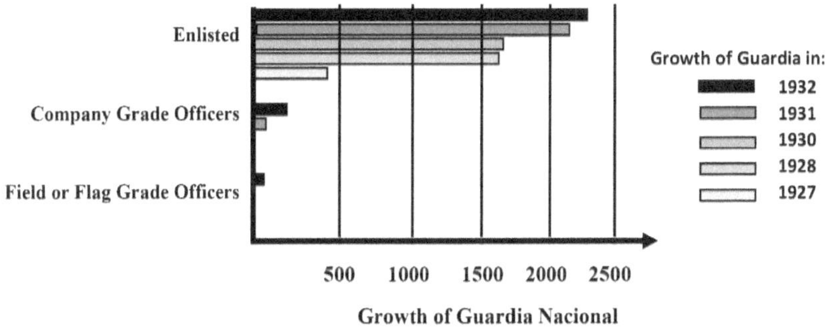

Growth of Guardia in:
1932
1931
1930
1928
1927

Growth of Guardia Nacional

A year later, and in preparation for assuming the duties of the 2nd Marine Brigade after the 1928 elections, the Guardia Nacional grew 400%. By the end of 1928, there were 1,637 Guardia. The Guardia Nacional remained at this manning level from 1928 to 1930. The Guardia Nacional de Nicaragua experienced its next substantial growth in enlisted troops in 1930–1931. During this time, the Guardia Nacional grew to 2,150. This was a response to US Secretary of State Henry L. Stimson's warning that Marines of the

2nd Brigade would be reduced in 1931 from 1,500 to only 500. In 1932, the Guardia Nacional de Nicaragua stood at 2,274 troops on the eve of the Marines withdrawal. However, this number does not count the 1,409 auxiliares, or reservists, who could be called up in an emergency.

From 1927 to 1932, in less than six years, the US Marines had created a security force largely from scratch and then transitioned complete responsibility for Nicaraguan security back to the host nation. From 1929 until the end of the intervention, the bulk of patrolling and operational responsibilities were transferred from US Marines to the Guardia Nacional de Nicaragua. By Spring 1930, less than two years after the start of the intervention, Guardia Nacional forces were aggressively patrolling in place of Marine units. Though there would be a few major Marine operations in times of emergency, the fight was largely taken over by the Guardia by 1930.

Even though they lacked specific training, the Marines had tangible successes in creating effective police. In American political culture, there is a clear distinction between military and policing duties. From the start of the intervention, Marines were well equipped to teach the basics of military tactics. However, they had little to no experience in training police officers. Immediately upon taking over the Managua Police, Marines realized that Nicaraguan police had even less experience than they did. Captain Herbert S. Keimling, the first Marine to assume the role of the Managua police chief, noted that Managua police did not know what laws were then in force.[54] Previously, the Managua police had simply arrested those whom they bore a grudge against or those who refused to pay bribes. Keimling relates that he set up policing schools to instruct police on what laws were in force and how to properly enforce these laws. As a result of his efforts, revenue from fines and arrests increased for legitimate offenses, and the total number of offenses began to drop.[55] The Marines also dealt with police corruption through strict oversight, routine formations, station visits, and reporting requirements.

The Tumbling State   153

Often missed in analyzing the construction of a modern military force is the combat-support and combat-service-support enablers, without whom modern warfare is near impossible. The Marines brought both combat support and combat-service support along with military technological innovations in the areas of medical, communications, and supply operations. An important result of the Marines' efforts was that the Guardia Nacional was able to sustain these adaptations after the Marines withdrew. Close air support was the only innovation the Marines employed in Nicaragua that the Guardia Nacional lacked the economic, technological, and industrial infrastructure to sustain after US withdrawal.

According to Millett, "the Marines laid the basis for the most efficient communications system Nicaragua had ever known. Unfortunately, they also left a heritage of military control over communications, a heritage which has continued up until the present."[56] The impetus for this military upgrade to Nicaragua's communication systems was the ease with which insurgents were able to cut the physical telegraph lines. Wireless communications were still a relatively nascent technology in the late 1920s, and yet the Marines established the first reliable transisthmian wireless communication system in September 1931.[57] The Marines also trained Nicaraguan communicators to run the network after the Marines withdrew.

In addition to communications, the Marines made a significant impact on the Guardia Nacional de Nicaragua's medical and logistical services. Overcoming Nicaraguan medical and logistical shortfalls occupied a significant portion of Marine attention immediately after arrival. Guardia Nacional personnel were constantly burdened by the continuous threat of syphilis, malaria, dysentery, and typhoid epidemics. To address these threats, the Marines of the Guardia Nacional initially secured the services of US Navy medical corps personnel. Later the Marines brought in American medical corps personnel to run local corpsmen courses and assist in the establishment of several military hospitals. These rudimentary services would most likely have been considered woefully inadequate by even US domestic standards of the time. They nonetheless represented a remarkable improvement over what was previously available to Nicaraguan troops at the time. Lastly, the Marines provided the Guardia Nacional with formal quartermaster regulations and modern transportation systems. These new systems professionalized the ability of the Guardia to sustain itself during extended combat operations.

By the time the Marines withdrew in January 1933, the Guardia would be so successful in its primary mission that it was also assigned a wide array of services beyond normal military and policing functions. The Marines and Guardia operated Nicaragua's prisons, sewage and waste functions, control of arms sales, liquor sales during elections, and humanitarian assistance operations. What is noteworthy is that, despite receiving little to no formal training in any of these areas, Marines nevertheless trained the Guardia Nacional to execute these tasks with aplomb for decades after the Marines departed.[58]

Attempts to politicize the Guardia Nacional began almost immediately during the first negotiations for the establishment of a nonpartisan security force in 1925. When Major Carter's constabulary was initially formed in 1925, its most passionate political enemy was Emiliano Chamorro. He saw a tactically effective Guardia as an impediment to revolution. He worked to delay passage of a bill recognizing and funding the force. Yet, once he gained power, he was all too content to invest in and support the development of the force. However, with the attrition the Guardia Nacional suffered before 1927, Chamorro was not able to reap significant benefit from his newfound support for the Guardia.

Prior to his election, President José María Moncada Tapia had professed a desire for not only a very capable, but also a very nonpartisan security force to secure the election. However, following his election in 1928, Moncada immediately began to try to politicize the force. In 1929, 1930, and 1931, President Moncada, the former insurgent, demanded the Guardia Nacional arrest his political opponents. When the Guardia resisted, he attempted to place the regional Guardia Nacional forces under the control of local *jefe politicos*, the political bosses, who were entirely partisan and politically reliable to Moncada. When the Marines of the Guardia Nacional resisted that effort, Moncada began to target the Guardia Nacional's budget. From 1930–1931, Moncada had some success at reducing the Guardia Nacional's budget. However, this success was obviated by increased Sandinista threat, which prevented the intended cuts.

Finally, Moncada attempted to create his own personal forces in the form of auxiliaries and civicos to offset the power of the nonpartisan Guardia. However, the Marines of the Guardia Nacional were successful in encouraging legislative support to keep these reserve forces nonpartisan. The Marines were also successful at bringing these forces under their training umbrella and authority.

Whereas Moncada had been prevented from turning the Guardia Nacional into his personal force, this nevertheless placed an inordinate amount of power in the hands of one man—the Jefe Director of the Guardia. The consolidation of such power in one office would be a significant problem once the Americans left. The Marines had laudable success in maintaining the nonpartisan nature of the Guardia Nacional while they were present. However, their failure to grow sufficient field and flag grade officers early enough would doom the Guardia Nacional initially to bipartisanship and ultimately into complete partisanship.

The Marines inexplicably waited until three years into the intervention to begin to formally develop a Nicaraguan officer corps. The Marines had appointed capable Guardia noncommissioned officers to commissioned officer ranks from the start through battlefield promotions. However, by the time the Marines departed, the Guardia Nacional could barely fill its requirement for company-grade officers and did not have a single professionally trained field-grade officer (see figure 7). According to Gravatt, "more than three-quarters of all academy officers graduated in 1932, the last year of the intervention. For field-grade officers, there was no training—formal or informal—whatsoever. There were no Nicaraguan officers above the rank of first lieutenant until November 1932."⁵⁹ It currently takes more than twenty-two years to produce an American general or flag-grade officer in the US military. The Marines responsible for developing the Guardia may have assumed they had more time to develop officers given the thirteen-year duration of their previous intervention. Nevertheless, they still would have been just beyond the cusp of growing even the lowest-ranked field-grade officers if they had begun their officer-training efforts immediately upon arrival.

Table 7. Growth of the Guardia Nacional, 1927–1932.

	1927	1928	1930	1931	1932
Field or Flag Grade Officers	0	0	0	0	13
Company Grade Officers	0	0	15	37	180
Enlisted	438	1,637	1,650	2,150	2,274

The primary means of creating officers in the Guardia Nacional was via the Academia Militare. The Academia had three Marine officers in charge of instruction. The Academia turned out nine cadets in its first class in June 1930. By September 1930, combining Academia graduates with officers who were formerly enlisted Guardia brought the total of Nicaraguan officers to fifteen out of a 220-officer cohort. The Guardia's officer corps began to increase dramatically with US Secretary of State Henry L. Stimson's warning that the US would withdraw 1,000 of the 1,500 Marines in Nicaragua in 1931. However, the challenge was still in finding capable officer candidates. Of the ninety-seven Nicaraguans who applied for the November 1930 class, only thirty-seven were accepted. The bulk of these officer candidates hailed from middle-class families because, as Millett argues, "the wealthy were uninterested and the poor were unqualified."⁶⁰ The Marines would also experience firsthand the power of partisanship in officer candidate selection. President Moncada had to personally approve each candidate to the Academia Militare. He therefore disproportionately approved applications from known liberal families over conservative ones.

Even with increased training of junior officers, American representatives began to realize by mid-1932 that there was not sufficient time remaining to train enough field-grade officers before the Marines departed. Additionally, Secretary Stimson would not permit Marine officers to remain in Nicaragua after 1933. General Matthews, the last Marine commander of the Guardia Nacional, suggested that both conservatives and liberals draw up lists of potential field-grade officers split evenly between the two parties. Each party's list had to have both liberal and conservative officers on them. Then, the outgoing president would appoint all the names from the winning candidate's list.

This plan to at least achieve a measure of bipartisanship did three key things. First, it surrendered any hope of a nonpartisan force. Second, because the names for these officers were aligned with the political elites, the primary qualifier for these officers would be political reliability rather than tactical capability. Third, once the Jefe Director was selected by the winning party, it was a forgone conclusion that he would gradually eject all the officers of the losing party and create an entirely partisan force.[61] As a result, initially, three of five colonels and six of eight majors selected were liberals—the winning party. The Marines only had two weeks to train these brand-new field-grade officers for their new roles.

In November 1932, the prominent liberal and former insurgent, President Juan Bautista Sacasa, was elected president. With the imminent withdrawal of the Marines, Sandino promised to end his fighting. This was a potential boon for a country torn apart by strife for over six years. Another boon, at least for the Liberal party, was that Sacasa would get to select the first Nicaraguan Jefe Director of the Guardia Nacional. With Sandino's promise, whether it was true or not, and the control of the Guardia Nacional in hands supportive of the current regime, there was reason to be optimistic. And yet, few were. In particular, Laurence

Duggan, head of the US State Department's Latin American Division, predicted "upon the withdrawal of the Marine officers in the Guardia next fall, the forces of disintegration will be set into action."[62]

The failure to create field-grade officers early enough led to the Guardia Nacional becoming bipartisan. The failure to develop a nonpartisan senior executive of the Guardia guaranteed that Guardia Nacional would quickly become entirely partisan. The only Guardia who were even superficially imbued with a requirement for nonpartisanship were the enlisted and the junior officers. However, they would have little influence in the Guardia. The field-grade officers selected near the end of the Marines' tenure were at least selected on a bipartisan basis, which temporarily afforded some measure of neutrality. However, the one office of the Guardia Nacional that was exempt of the need for nonpartisanship or even bipartisanship was the office of Jefe Director. This was chosen entirely by political loyalty in hopes that the incoming president would at least have the support of the military and the police to govern the state.[63] The Americans who worked with Anastasio Somoza preferred him to the other candidates, General Jose Maria Zelaya and General Gustavo Abaunza. The liberals had little reason to protest Somoza's selection as he was a devoted liberal and former insurgent who had originally fought against the Conservatives and Americans. However, Somoza's personal ambition and the manner in which Sandino was welcomed back into the government set the conditions that would eventually lead to Somoza's reign.

The First Test—Defeat of the Sandinistas

The deal that Sandino negotiated with President Sacasa at midnight on February 2, 1933, for his surrender and the end of the insurgency was a very generous but dangerous one. Whereas the negotiations held out the promise of peace, they also aggravated Somoza and the Guardia who had lost comrades who were killed by the Sandinistas. In contravention of the Stimson Peace Plan or Pact of Espino Negro, Sandino was allowed to maintain a private militia of 100 men who would act as his personal security force. The former insurgents would control territory ceded to Sandino in Nuevo Segovia and be allowed to keep their weapons. More disturbingly, the government would also have to bear the cost of their continued service as Sandino's personal army. Furthermore, the Nicaraguan government would have to provide contracts for public works projects in Sandino's territory for which only Sandino loyalists could compete.[64] In return, Sandino would completely disarm the rest of his forces and submit to government control.

These accommodations made for the Sandinistas were onerous to the Guardia's personnel, in general, and to Somoza, in particular. Yet, it was Sandino's refusal to disarm according to the agreement and Sacasa's decision to consider elevating a former Sandinista over the Guardia that precipitated Somoza's assassination of Sandino. Whereas Sandino's forces did turn in some of their weapons, Sandino kept far more than 100 armed men and weapons. In less than a year after his supposed disarmament, he offered the services of his 600 Sandinistas to the exclusive service of President Sacasa.[65]

Somoza demanded that Sandino surrender his weapons on February 4, 1934, in accordance with the surrender agreement of February 2, 1933. However, not only did Sandino refuse to abide by the agreement, he convinced President Sacasa to appoint a presidential delegate to govern Nicaragua's northern provinces. This delegate would "protect the interests of Sandino and his followers" and thereby place the Guardia in these provinces under a former insurgent officer and under Sandino's direct control.[66] This would have been analogous to (and about as popular as) President Lincoln appointing Robert E. Lee to command Union forces after the US Civil War.

Unsurprisingly, the situation came to a head when President Sacasa held a party on behalf of Sandino and Sandino's former insurgents on February 21, 1934. Immediately following the party, Somoza and his officers arrested Sandino and had him and his officers executed and buried in unmarked graves.[67] The following day, Sandino's forces encamped at Wiwili, Nicaragua, were given the opportunity to disarm as required by the 1933 surrender agreement. When the Sandinistas refused, the entire camp was wiped out, killing between 22–300 Sandinistas, depending on differing accounts.[68] Somoza and the Guardia Nacional were able to do in two days what had previously been impossible for five years. Even as there would be some half-hearted attempts to reconstitute the Sandinista movement, no real threat to the Nicaraguan government would present itself again for another forty-six years. Within a month of Sandino's execution, Sacasa would begin replacing Guardia officers with his own relatives. Within two years, Somoza would pressure Sacasa to resign, and Somoza's family would begin its four-decade dynastic rule.

CONCLUSION

From 1927 to 1933, the US employed encadrement, an inhabiting strategy, to develop Nicaraguan security institutions. US Marines sought to effect evolutionary change in a revolutionary timeframe and did so from within Nicaraguan security institutions. With the exception of their development of a Nicaraguan officer corps, they were successful. Conversely, the US Department of State employed a strategy of coercive influence to advise the Nicaraguan government. US diplomats sought to effect evolutionary change in a revolutionary timeframe and did this from outside the Nicaraguan institutions and failed entirely. The combination of inhabiting and influencing strategies produced an environment of high stability but faux democracy. The combination of high stability and faux democracy describes what I delineate as a tumbling state. The Nicaraguan state, supported by the US-trained Guardia Nacional, was able to persist for forty-six years after US withdrawal.

In analyzing US efforts to develop the Nicaraguan government and its security forces, four outcomes are clear. First, the US succeeded in securing long-term Nicaraguan stability. Second, the US succeeded in creating the highest caliber military force possible. Third, the US failed to create a nonpartisan security force. Fourth, the US failed to change the political culture that had previously politicized the military and police of Nicaragua. The first visible observation that the US succeeded in maintaining stability in Nicaragua is evident in the reduced frequency of what had become perpetual armed revolt on an average of every three years.

The second observation is that the intervention produced a Nicaraguan security force without parallel in Nicaragua to that time. Millett notes that "despite such handicaps as well as the limitations placed on training time imposed by the demands of the campaign against Sandino, the Marines did surprisingly well, transforming their recruits within a few months into the best trained, disciplined, and equipped force in Nicaraguan history."[69] Whether or not their tactical skills and discipline would be used to noble ends, there is great support for the argument that the Marines created the most tactically and technically capable military force Nicaragua had ever possessed to that date.[70]

The third observation is that the US failed to produce the nonpartisan security force the US had thought was the solution to Nicaraguan instability. As Millett argues, "had the achievement of technical and organizational improvements been the sole objective, the Guardia's creation would have been rated, with some qualifications, a major success. However, the Guardia did not completely defeat the Sandinistas from 1927 to 1932, and the Guardia failed to become a non-political force."[71] Just as there is great support for the unrivaled military success of the Guardia Nacional, there is similar consensus on the failure of the US to imbue this force with an enduring nonpartisan culture.[72]

The fourth and final observation is that the US failed to change the oligarchical kleptocratic political culture of Nicaragua during its tenure. As Gravatt advances, "the United States was not successful in changing Nicaraguan politics where continuance in the presidency or accession thereto was accomplished by force rather than by vote. The situation in post-intervention Nicaragua, Somoza and the National Guard exemplifies the difficulty of imposing permanent change from without."[73] Moreover, whereas Walker and Wade speak positively of the Frente Sandinista–led rule from 1979 to1990, they also acknowledge that from 1821 to 1979, at least "throughout␣Nicaraguan history, a small elite controlled most of the means of production and garnered most of the benefits. The country's rulers—whether openly dictatorial or ostensibly Democratic—almost always governed on behalf of the privileged few."[74] There does not seem to have been any great difference between liberals and conservatives when it came to graft, corruption, or perpetuation of power or seizure of power through force. In fact, at different times, the US helped perpetuate both parties for lengthy periods of time in office. It is this failure to change the oligarchical kleptocratic political culture that underwrites the other US failure—failure to create a nonpartisan security force.

What this analysis means in a larger sense is that no amount of military success can overcome a corrupt political culture. Another essential learning point in developing host-nation security forces is starting officer development as soon as is practicable. Thus, even if the intervention is of short duration, the training force must begin early to identify young leaders who might possess executive skills. These could be promoted expeditiously to prevent the need for appointments based on political reliability. However, even if the Marines in Nicaragua had begun earlier, it is inconceivable that a handful of neutral Guardia officers could remain perpetually immune from being corrupted by a universally oligarchical kleptocracy. Failure in the political culture is failure in the entire system.

Notes

1. Richard Millett, *Guardians of the Dynasty* (Maryknoll, NJ: Orbis, 1977), 183.
2. The number of civilians killed or wounded was not scrupulously recorded by either side. For the Marines all contacts and casualties are recorded in *The Official List of Contacts of the Guardia Nationale de Nicaragua, Headquarters Guardia National (1927–1933)*, Managua, Nicaragua: Division of Operations and Intelligence, 1933) from Marine Corps Archives, Julian C. Smith Collection, RG 202, Box 2, Folder 8, hereafter referred to as *The Official List of Contacts of the Guardia Nationale de Nicaragua*.
3. Richard Millett, *Guardians of the Dynasty: A History of How the U.S Created Guardia Nacional De Nicaragua and the Somoza Family* (Ossining, New York: Orbis Books, 1978), 139.
4. From the end of Sandino's fight February 2, 1933 until the *Frente Sandinista* capture of Managua on July 17, 1979.
5. Gravatt, *The Marines and the Guardia*, 273.
6. Ibid.
7. Millett, *Guardians of the Dynasty*, 42.
8. "United States Intervention, 1909–33," Library of Congress (1993), http://countrystudies.us/nicaragua/10.htm.
9. Gravatt, *The Marines and the Guardia*, 3–4; Richard Millett, *Searching for Stability: The US Development of Constabulary Forces in Latin America and the Philippines* (Fort Leavenworth, Kansas: US Army Combined Arms Center-Combat Studies Institute Press, 2010).
10. Charles Neimeyer, "Combat in Nicaragua," (April 2008), https://www.mca-marines.org/gazette/2008/04/combat-nicaragua.
11. Neimeyer, "Combat in Nicaragua."
12. *The Official List of Contacts of the Guardia Nationale de Nicaragua*, Report #1.
13. The roundness of these numbers call into question the degree of their precision. However, even as they are merely estimates, they are indicative of the Marines and Guardia getting the better of the initial conventional fights and this aligns well with Sandino's switch to guerilla operations.
14. *The Official List of Contacts of the Guardia Nationale de Nicaragua*, 2–21.

15. General Julian C. Smith, *Area Order 3: Martial Law in the Central Area* (March 1931), USMC Archives, Julian C. Smith Collection, RG 202, Box 2, Folder 10.
16. Ibid.
17. Thomas Walker and Christine Wade, *Nicaragua: Living in the Shadow of the Eagle* (Boulder, CO: Westview Press, 2011), 2–3.
18. Walker and Wade, *Nicaragua*, 3.
19. "Nicaragua: Historical Demographical Data of the Whole Country," http://www.populstat.info/Americas/nicaragc.htm: 1912—562k, 1916—600k, 1923—652k, 1927—670k, 1933—697k.
20. Walker and Wade, *Nicaragua*, 2.
21. Combination of Spanish and Indian forebears.
22. I leverage terminology from the USAID tactical conflict analysis framework here. For more on this see http://pdf.usaid.gov/pdf_docs/Pnadn621.pdf.
23. Library of Congress, "United States Intervention, 1909–33."
24. Millett, *Guardians of the Dynasty*, 22–23; James T. Wall, *Manifest Destiny Denied: America's First Intervention in Nicaragua* (Washington, DC: University Press of America, 1961), xvi.
25. Millett, *Guardians of the Dynasty*, 98.
26. Smith, *Area Order 3*.
27. Millett, *Guardians of the Dynasty*, 91.
28. Gravatt, *The Marines and the Guardia*, 158; Neimeyer, "Combat in Nicaragua."
29. Dieter Nohlen, ed., *Elections in the Americas: A Data Handbook*, vol. 1 (Oxford: Oxford University Press, 2005), 500; and also http://www.populstat.info/Americas/nicaragc.htm.
30. Krepinevich, *The Army and Vietnam*, 236; Linn, *The Philippine War*, 325.
31. 6,700 troops to cover approximately 50,338 square miles of territory.
32. Gravatt, *The Marines and the Guardia*, 157; *Memorandum for General Matthews, Guardian Nacional de Nicaragua, 2d Marine Brigade*—October 1932, USMC Archives, Julian C. Smith Collection, RG 202, Box 2, Folder 12; *The Official List of Contacts of the Guardia Nationale de Nicaragua*; Millett, *Guardians of the Dynasty*, 148.
33. Based on an estimate of 3,000 Sandinistas.
34. Neimeyer, "Combat in Nicaragua."
35. *The Official List of Contacts of the Guardia Nationale de Nicaragua*; Neimeyer, "Combat in Nicaragua."
36. Neimeyer, "Combat in Nicaragua."

37. Millet, *Guardians of the Dynasty*, 62–63.
38. From the Julian C. Smith Collection as captured in Millett, *Guardians of the Dynasty*, 62 and 71.
39. Millett, *Guardians of the Dynasty*, 104–108; Neimeyer, "Combat in Nicaragua."
40. Gravatt, *The Marines and the Guardia*, 102–123; Millett, *Guardians of the Dynasty*, 114–116, 149; Bernard C. Nalty, *The United States Marines in Nicaragua* (Washington, DC: Historical Branch, Headquarters USMC, 1968); Neimeyer, "Combat in Nicaragua."
41. Dana Carleton Munro, *The United States and the Caribbean Republics, 1921–1933*, (Princeton: Princeton University Press, 1974), 252.
42. Boot, *Savage Wars of Peace*, 243; Millett, *Guardians of the Dynasty*, 106.
43. Nohlen, *Elections in the Americas*.
44. Millett, *Guardians of the Dynasty*, 30–31.
45. Gravatt, *The Marines and the Guardia*, 30–36; Millett, *Guardians of the Dynasty*, 42; Neimeyer, "Combat in Nicaragua."
46. Millet, *Guardians of the Dynasty*, 34–35.
47. Ibid., 48.
48. Gravatt, *The Marines and the Guardia*, 158.
49. Millett, *Guardians of the Dynasty*, 21.
50. Ibid.
51. Millett, *Guardians of the Dynasty*, 22.
52. Ibid.
53. A portion of which were holdovers from Major Carter's Constabulary who had to be discharged first and then screened and re-enlisted.
54. H. S. Keimling, "Report on Policing in Managua: March 11, 1933," in *Papers of the Jefe Director of the Guardia Nacional de Nicaragua*, National Archives RG 127, Item 198.
55. Ibid.
56. Millett, *Guardians of the Dynasty*, 76.
57. Ibid.
58. Boot, *Savage Wars of Peace*, 248; Gravatt, *The Marines and the Guardia*, 137–155; Millett, *Guardians of the Dynasty*, 71–79; Nalty, *United States Marines in Nicaragua*.
59. Gravatt, *The Marines and the Guardia*, 3.
60. Millett, *Guardians of the Dynasty*, 126.
61. Gravatt, *The Marines and the Guardia*; Millett, *Guardians of the Dynasty*; Nalty, *United States Marines in Nicaragua*.

62. Andrew Crawley, *Somoza and Roosevelt: Good Neighbor Diplomacy in Nicaragua, 1933–1945* (Oxford: Oxford University Press, 2007), 22.
63. Munro, *United States and the Caribbean Republics*.
64. Crawley, *Somoza and Roosevelt*, 30–32.
65. Ibid., 45.
66. Millett, *Guardians of the Dynasty*, 155.
67. Crawley, *Somoza and Roosevelt*, 46.
68. Accounts also vary as to whether or not women and children were involved as well; Gravatt, *The Marines and the Guardia*; Millett (1978), 159; Nalty, *United States Marines in Nicaragua*.
69. Millett, *Guardians of the Dynasty*, 71.
70. Boot, *Savage Wars of Peace*, 246–247 and 250–251; Gravatt, *The Marines and the Guardia*, 154–155; Library of Congress, "United States Intervention, 1909–33"; Munro, *United States and the Caribbean Republics*; Nalty, *United States Marines in Nicaragua*; Neimeyer, "Combat in Nicaragua"; Walker and Wade, *Nicaragua*, 196.
71. Millett, *Guardians of the Dynasty*, 79.
72. Boot, *Savage Wars of Peace*, 250; Gravatt, *The Marines and the Guardia*, 155; Library of Congress, "United States Intervention, 1909–33"; Munro, *United States and the Caribbean Republics*; Nalty, *United States Marines in Nicaragua*; Neimeyer, "Combat in Nicaragua"; Walker and Wade, *Nicaragua*, 22.
73. Gravatt, *The Marines and the Guardia*, 4.
74. Walker and Wade, *Nicaragua*, 3.

CHAPTER 5

THE CRUMBLING STATE

VIETNAM

The US Congress passed the Gulf of Tonkin Resolution on August 7, 1964, providing legislative consent to expand US involvement in Vietnam. In September 1964, the Director of Central Intelligence published the Special National Intelligence Estimate: Chances for a Stable Government in South Vietnam (SNIE 53-64).[1] Delivered five months prior to the insertion of the first US combat formations, the first two sentences of SNIE 53-64 were prescient. The first stated, "THE PROBLEM-To assess the chances for the emergence of a stable non-Communist regime in South Vietnam." The second declared, "CONCLUSION-At present the odds are against the emergence of a stable government capable of effectively prosecuting the war in South Vietnam." [2]

American policymakers became convinced in 1964 that only increasing amounts of military aid to South Vietnam coupled with US troop intervention could facilitate the "emergence of a stable government capable of effectively prosecuting the war in South Vietnam." By April 1975, the US had invested $133.38 billion in economic aid[3] and $519.48 billion in military aid (in 2019 dollars).[4] This included an estimated $20.8 billion in lost or destroyed US equipment and as much as $5.21 billion in equipment abandoned by the South Vietnamese to the North Vietnamese.[5] In addition to these colossal expenditures in treasure, the cost in terms of blood was also considerable. In addition to countless North and South Vietnamese lives lost, the US lost 58,220 of its own citizens in the war.[6] In the last months of the war, there was little sign that the perpetual payment of these rents would cease. In FY1974, the US provided $3.32 billion in economic assistance[7] and anticipated paying an additional $2.3 billion[8] in FY1975 (in 2019 dollars).[9]

Despite all this, the prognosis at the end of the war for a "stable government capable of effectively prosecuting the war in South Vietnam" was no better than at the beginning. On May 23, 1974, a decade after SNIE 53-64, the CIA published another analysis of the situation facing the Army of the Republic of Vietnam (ARVN) and the Government of (South) Vietnam (GVN) immediately preceding the final North Vietnamese offensive. It found that, "Should a major offensive occur...ARVN might be unable to regain the initiative, and it would be questionable whether the GVN would be able to survive without combat participation by US Air Force and Navy units. At a minimum, large-scale US logistic support would be required to stop the communist drive."[10]

Examining the Case—Vietnam

> No level of operational or tactical proficiency can overcome corrupt host-nation governance and national military leadership.
> —General Anthony Zinni (USMC, ret)[11]

In 1965, Senate Majority Leader Mike Mansfield visited Vietnam to observe the situation there firsthand, and he returned from his trip with two assessments. First, he noted that the $2 billion the US had spent to develop the South Vietnamese government and the Republic of Vietnam Armed Forces (RVNAF) since 1954 had been wasted. Second, he assessed that the only way to arrest the deteriorating situation in South Vietnam was either through the "massive commitment of US forces" or through nuclear war.[12]

On March 8, 1965, US Marines of Battalion Landing Team 3/9 waded ashore at China Beach near Danang. These Marines represented the first combat formations to arrive in Vietnam. A few months later, elements of the 101st Airborne arrived. These forces were the first of fifty-four allied maneuver battalions requested by General William Westmoreland.

Westmoreland, the commander of Military Assistance Command Vietnam (MACV), asserted that these forces were the only means to halt a potential collapse of the Government of South Vietnam and the RVNAF. At its height, US troop levels in Vietnam were greater than the peak troop levels of all three other case studies of this analysis combined. Three key factors led to the decision to intervene in such a drastic fashion after over a decade of support for the Government of South Vietnam. The first was the burgeoning strength of the Viet Cong main force units.[13] The second was the comparative weakness of the RVNAF. And the third was the fragility of the South Vietnamese government. All these took place simultaneously within the context of the Cold War.

On December 29, 1960, US forces in Vietnam observed firsthand the transition of what Mao would term a Phase I guerilla force into a Phase II force.[14] On that day, the communist state of North Vietnam publicly announced the formation of the National Liberation Front (NLF) in Hanoi. Since the year prior, the Viet Cong (VC), the military wing of the NLF, had been establishing an extensive logistical network. Eventually, the logistical network, known as the Ho Chi Minh Trail, would consist of trails and roads from North Vietnam through parts of Laos, Cambodia, and South Vietnam. Through the efforts of Group 559, delivery of VC troops and materials began by August 1959. In their first major engagement, 400 VC would do poorly against the ARVN forces at Kien Hoa in April 1961. However, VC forces would increase rapidly during 1961 and would grow to 17,000 troops, with 89 percent of these forces locally recruited and supplied from South Vietnam. Despite their initial setback in Kien

Hoa, by the end of 1961, the VC would control a substantial portion of the Mekong River Delta.[15] By January 1963, the VC had grown enough in capacity to defeat ARVN forces at Ap Bac. The growth and success of the VC was not enough, by itself, to convince US policymakers to intervene in Vietnam with massive combat formations.

The anemic capacity of the South Vietnamese government also contributed to the requirement for the intervention of US combat formations. The implosion of the government of South Vietnam, after the anti-Diem coup in 1963, did not come as a surprise to US policymakers.[16] President Diem had been an expat during the years of the Vietnamese nationalist battles against the French. Once back in Vietnam, Diem rejected both Chinese and Soviet communism and western colonialism. To make matters worse, he also lacked a significant constituency beyond the wealthy Vietnamese Catholics. The fear of Buddhist cooption by the Communists caused the Diem regime to be particularly repressive toward Buddhists and political rivals and increasingly distrustful of US agents. By November 1963, the US was aware of and assented to a final coup attempt against Diem.[17] President Diem was deposed by General Nguyen Khanh and eventually murdered in November 1963. Brigadier General James Lawton Collins Jr. asserted that the combination of Diem's repression of the Buddhists and Diem's subsequent overthrow and murder "led to a complete deterioration of the armed forces. Only the decision to introduce US combat forces in early 1965 saved the Republic of Vietnam from total military defeat."[18]

COIN Methods: Search and Destroy and The Other War

> When General Westmoreland was asked at a press conference what the answer to insurgency was, his reply was one word: 'Firepower.'
>
> –Andrew Krepinevich, *The Army in Vietnam*[19]

Both enemy- and population-centric COIN methods were present, weighted and effective during the US intervention in Vietnam from 1965 to 1972. The US intervention in Vietnam can be divided into two major strategic divisions. First was the search-and-destroy period from 1965 to 1968 where the US weighted enemy-centric COIN methods to destroy the VC. The second period is from 1968 to 1972 where the US weighted population-centric COIN methods to win the support of the South Vietnamese populace. It is important to note the use of the term *weighted*. The use of this term is designed to demonstrate that at no time did US forces completely forgo either COIN method entirely. Rather, US efforts tended to focus more on one at the expense of the other at different points during the intervention.

In 1964–1968, the United States pursued two divergent strategies to develop a stable and capable South Vietnamese government. On June 20, 1964, Westmoreland assumed command of the MACV. Operating in near simultaneity, General Westmoreland pursued the destruction of the insurgency, while Robert Komer, the head of the Civil Operations and Revolutionary Development, or CORDS program, pursued "combined political military efforts to defeat the rural insurgency." Despite their near simultaneity, Westmoreland's access to the massive American military capacity present in Vietnam allowed his strategic thumbprint to overshadow Komer's until after the Tet Offensive in January 1968.

Westmoreland envisioned the destruction of the Viet Cong and the defense of South Vietnam against North Vietnamese invasion as the optimal means to achieve US goals in Vietnam. Westmoreland saw the VC and NVA main force units as bullies with crowbars who could tear down the South Vietnamese "house" far more rapidly than the insurgent "termites." Westmoreland's logic was founded upon the idea that while both conventional main force units and insurgents could tear down the GVN's "house," the former could do it considerably faster than the latter.[20] His opinion is understandable given the defeat of the ARVN at Phouc Long and Quang Ngai and what he saw as the "tactical stupidity and cowardice" of ARVN officers fighting the VC in 1963.[21] The ability of the VC main force units to defeat conventional ARVN battalions conformed well to the pattern previously established by the Viet Minh, the forerunners of the VC, against the French in 1954. Resultantly, Westmoreland abandoned his 1964 request for US COIN experts in favor of requesting air-mobility experts being sent to Vietnam.[22] In his estimation, the Viet Cong main force threat could be defeated most effectively and rapidly by leveraging considerable US advantages in tactical mobility and firepower.

Early major heliborne operations by US Army and Marine forces seemed to confirm Westmoreland's appraisal of the situation. US forces had substantial early successes in Operation Starlight (August 17–24, 1965), Operation Long Reach (October–November 1965), and Operation Harvest Moon (December 8–20, 1965). The VC main forces had difficulty adjusting to the vast mobility advantage possessed by US forces and were easily isolated and then bludgeoned by US firepower. Initially, NVA and VC main force commanders had "concluded the best way forward was to crush the Americans in open battle..."[23] However, after being punished by US conventional advantages, the VC and NVA main forces began to deny US forces large, lucrative, static targets. VC forces would thereafter only give battle when they possessed a distinct local advantage.

From 1965–1967, Westmoreland pursued an elusive crossover point in the cumulative attrition of the VC and NVA forces. Westmoreland sought to attrite the communists' ability to continue to fight. He worked to reach this crossover point by destroying enough Communist forces and material that they would be unable to recoup losses at a rate sufficient to remain effective. In pursuit of this crossover point, US forces conducted numerous large-scale search-and-destroy operations. Some of the larger operations include Operation Van Buren (Jan 1966), Operation Masher (Jan–Mar 1966), Operation Hastings (Jul–Aug 1966), Operation Cedar Falls (Jan 1967), Operation Junction City (Feb–May 1967), and Operation Attleboro (Sept–Nov 1967).

The Crumbling State

By April 1967, General Westmoreland, largely based on inflated body counts and saturation bombing by the US Air Force, was convinced that the war was near the crossover point—at least with respect to the destruction of the VC. His assessment would be invalidated by the Tet Offensive that took place eight months later.

In January 1968, General Vo Nguyen Giap commenced the Tet Offensive, a blanket offensive by all VC forces in South Vietnam. Giap had erroneously assessed the Government of South Vietnam's center of gravity as its connection with the Vietnamese people.[24] Giap had been successful at drawing some, though not most, of the largest US units away from the populated coastal regions. This left these more populated coastal areas less secure than if the larger US units had not been drawn into the less populated hinterland.

With the launch of the offensive, VC main force units seized Hue City, surrounded the US Marine Base at Khe Sanh, and stormed the US Embassy in Saigon. From General Westmoreland's perspective, this was the large-scale set piece battle he had sought since the early successes of 1965. Despite causing significant American and ARVN casualties and embarrassment, the NVA lost Hue City and were not able to overrun Khe Sanh. More critically, as a result of the Tet Offensive, the VC were largely obliterated as an effective fighting force. Still, with the offensive taking place after Westmoreland had predicted the achievement of an erroneous crossover point in 1967, the US war effort in Vietnam was the real strategic loser. The Tet Offensive would result in a seismic shift in US Vietnam strategy and a change in MACV command from Westmoreland to General Creighton Abrams on July 1, 1968.

Intriguingly though, in the same month that MACV approached its erroneous crossover point, Robert Komer alternately saw the need to redirect the war effort in a different direction. In April 1967, Komer asserted that in place of destruction of the VC and NVA, the US effort should shift its focus toward the pacification of the rural population. CORDS and programs like the Marine Corps' Combined Action Platoons

(CAP) constituted sideshows from 1964–1968 to the US Army's paramount focus on attrition. These sideshows became disparagingly known as "The Other War," or the war that diluted resources from the Westmoreland's preferred war of attrition.

The advent of the CORDS pacification program in May 1967 represented the second arm in the bifurcated Vietnam strategy from 1964 to 1968. CORDS has been regarded as "a successful integration of civilian and military efforts" to combat the South Vietnamese insurgency.[25] The CORDS program's efficacy as a pacification and population-centric COIN method is widely attested to.[26] The principal negative critiques of CORDS are, at the worst, that gains were illusory.[27] More commonly, CORDS has been seen as being non-prioritized and under-resourced, incorporated too-late, or inhibited by corrupt South Vietnamese officials.[28]

After Abrams took over, the focus of US efforts shifted from the defeat or destruction of the VC to gaining the support of the South Vietnamese populace and providing for their security. Population-centric COIN methods were begun under Westmoreland but became the focus of effort under Abrams. These methods included CORDS, the District Advisory Program, the Phoenix Program, and the use of South Vietnamese forces to hold pacified areas. US combat formations would continue to do battle in the most dangerous areas in order to provide breathing space for the growth and strengthening of the RVNAF forces.

The accomplishments of Abrams' population-centric methods are difficult to argue with. In contrast to efforts before 1967, "by early 1970, 93 percent of South Vietnamese lived in "relatively secure" villages, an increase of almost 20 percent from the middle of 1968, the year marred by the Tet Offensive."[29] CORDS had been extended to all forty-four provinces of South Vietnam, and the communist insurgency was substantially weakened. CORDS combined political development with security development at the national and provincial levels. At the district and hamlet levels, political and security development was effected through the District Advisory Program. And at the provincial level, the controversial Phoenix Program eviscerated the Viet Cong Infrastructure (or VCI).[30]

COIN THEORIES OBSERVED

American policymakers and historians could reasonably argue that the US had both defeated the insurgency militarily by 1968 (the VC main force units) and gained effective control of the South Vietnamese population by 1972. What they could not claim is that the South Vietnamese government would be able to last more than three years after the last US combat formations departed in August 1972. American commanders could reasonably argue that they had succeeded in both enemy- and population-centric COIN methods. The VC ceased to be militarily relevant after the Tet Offensive of 1968.[31] Tactically, Westmoreland did not arrive at his projected crossover point[32] in April 1967. He did arrive at this crossover point in July 1968—at least in reference to the VC.

Additionally, whereas Westmoreland is most commonly associated with enemy-centric COIN and Abrams with a more population-centric method, CORDS had its genesis under Westmoreland and its payoff under Abrams. The effects of CORDS operations from 1967 to 1972 are manifest. The Hamlet Evaluation System (HES) claimed 93 percent pacification of all South Vietnamese villages by 1970 and much reduced levels of insurgency in all forty-four provinces.[33] Similarly, the efficacy of the Phoenix Program and its targeted destruction of the VCI is attested to by Mark Moyar[34] and Colonel Andrew Finlayson (USMC, ret),[35] a CIA advisor to the Provincial Reconnaissance Unit (PRU) in Tay Ninh. Though the HES's quantitative claims of pacification may be viewed critically,[36] the material support of the South Vietnamese population was not instrumental in the North Vietnamese invasions of 1972 or 1975. Unlike the French at Dien Bien Phu, it was not Viet Cong insurgents or guerillas who captured Saigon in April 1975. It was the conventional army of an adjacent state actor.

Troops Ratios

Troop ratios also fail to explain the rapid demise of the South Vietnamese government and RVNAF after the withdrawal of US combat forces. Indeed, even before the landing of the first US maneuver battalions in March 1968, the RVNAF and US advisors had almost met the most conservative threshold for troop to population requirements. At every point in the American-led portion of the war in 1965–1972, the combined US and RVNAF troop levels well exceeded even the most demanding requirements for troop-to-population ratios. Additionally, the US and RVNAF always maintained a significant quantitative advantage over the VC/NVA forces in South Vietnam. Despite these perpetual quantitative advantages from the start, the US was nevertheless forced to intervene in 1965 to prevent South Vietnamese failure, the VC were able to conduct a nationwide offensive in 1968, and the NVA were able to overrun the entire country in 1975.

Contemporary estimates of required counterinsurgent ratios to population range between 1:50 as the most conservative and 1:91 as the most liberal. Before the first US Marine battalions landed in 1965, the combined US and RVNAF forces already present in South Vietnam resulted in a 1:32 counterinsurgent-to-population ratio—significantly better than the even the most conservative requirement (see table 8). Even if this ratio is adjusted for what Krepinevich and Linn describe as foxhole strength, there remains a ratio of 1:53 counterinsurgent-to-population ratio able to engage directly with the populace and insurgents. This is nearly double the most liberal contemporary minimum estimate. Therefore, if Quinlivan's, McGrath's, and Brown's assessments are accurate, then the US should not have had to intervene with any additional forces after 1964.

The counterinsurgent-to-population ratio, already theoretically sufficient even before the arrival of US combat formations in 1964, would further balloon to more than five times greater than even the most conservative estimate of troops required for successful COIN operations. And yet there would be no concomitant improvement in population security.

The Crumbling State

By the advent of the Tet Offensive and the peak deployment of US troops during the war, the US would have only the most tenuous control over the majority of South Vietnamese rural villages and hamlets.[37] In 1968, the RVNAF also had more than 619,000 troops who were solely devoted to pacification. At this high point of US intervention, there was one US or Vietnamese counterinsurgent for every twenty South Vietnamese citizens or one for every thirty-three if foxhole strength is used. This is significantly less population for every counterinsurgent to control and protect than even the most conservative estimate requires. And this impressive population-to-counterinsurgent ratio would continue to increase after 1968 when US troop levels began to decline. However, this was not the only concern.

South Vietnamese officials and US strategists late in the war were more concerned with the ratio of US and RVNAF maneuver battalions to VC/NVA maneuver battalions in South Vietnam. According to an assessment provided to the National Security Council by the CIA in 1972, the RVNAF had a 4.8:1 maneuver battalion advantage over the NVA and VC forces operating in South Vietnam (table 8). This nearly five-to-one ratio was sufficient to defeat the NVA/VC Ester Offensive in March 1972—but only when supported by US logistics and air power. By 1975, the RVNAF possessed more than one million troops, almost the same as in 1972. The difference in 1975 was that the US was no longer supporting with logistics, close air support, or advisors. Therefore, the withholding of US logistics, tactical sir support, and advisors provides a better explanation for the failure of the Government of South Vietnam and RVNAF in 1975 than do troop ratios.

Table 8. Troop Ratios to Population, Terrain, and Adversary.

Year	US Troop Levels	RVNAF	US & RVNAF (60% Foxhole Strength)	Pop[c]: Troop	60% Foxhole Strength: Pop	Troop: Terrain	Foxhole Strength: Terrain	VC/NVA in South Vietnam	Maneuver Battalion Ratio (US & GVN: VC/NVA)
1964	233,000[A]	514k[B]	537,300 (322,380)	32:1	53:1	8:1m[2B]	5:1m[2B]	Unknown	Unknown
1966	385,300[A]	736k	1,121,300 (672,780)	15:1	25:1	17:1m[2B]	10:1m[2B]	163-183k & 260-280k[E]	
1968	536,100[A]	619k[B]	855,100 (513,060)	20:1	33:1	13:1m[2B]	8:1m[2B]	500-600k total Main Force 76,585[G]	2.6:1[J]
1972	24,200[A]	1 mil+[B]	1,024,200 (614,520)	17:1	28:1	16:1m[2B]	9:1m[2B]	145-165k[H]	4.8:1[J]
1974	--							182k[I]	

Source. Table compiled by author from a variety of sources.[38]

ENDURING INSURGENT SANCTUARIES

General Anthony Zinni and Max Boot have argued that the failure of the Government of South Vietnam and RVNAF in 1975 was not principally due to COIN methods or troop ratios.[39] They argue that the 1975 failure is best explained by two key factors: 1) the inability of the US and the Government of South Vietnam to foreclose on adversary sanctuaries throughout the war, and 2) the cessation of US support to the South Vietnamese in 1974–1975. There seems to be considerable evidence that the termination of US support to the South Vietnamese government had a decisive effect on the ability of the RVNAF to defend South Vietnam. However, the enduring presence of insurgent sanctuaries does not provide as satisfying an explanation for the rapid demise of the Government of South Vietnam and the RVNAF after withdrawal of US forces.

The presence of insurgent staging bases and sanctuaries in North Vietnam, Cambodia, and Laos do not explain the failure of the RVNAF in major conventional war in 1975. Harry Summers argued that the US

The Crumbling State 179

Army in Vietnam had been too focused on COIN and not focused enough on defeating the North Vietnamese conventional forces.[40] Having failed to achieve its ends through popular uprising in 1961–1968, the North Vietnamese government began infiltrating larger numbers of conventional combat formations into South Vietnam. As such, Summers advocated for Westmoreland's OPLAN El Paso which would have created an extended DMZ across northern South Vietnam and into the neighboring states to cut off the North Vietnamese access to South Vietnam and support for the VC.[41] CIA estimates had tracked the infiltration rates from North to South Vietnam from January to May 1966 with a high of 9,350 incursions per month to a low of 210.[42] Westmoreland reasoned that if the US could have isolated the insurgents in South Vietnam from their support in North Vietnam, US and South Vietnamese forces could have starved them out and defeated them in detail.

However, OPLAN El Paso was predicated on the erroneous assumption that the bulk of the VC support before 1968 had originated from North Vietnam. The assumption that the VC could have been defeated by cutting them off from North Vietnamese support was invalidated, however, by another CIA report, also in 1966. Its analysis estimated that at least 50% of VC logistics were sourced from South Vietnam and not transported from North Vietnam.[43] Krepinevich's and Boot's research also support the assertion that most of the VC logistical support, before 1968, originated locally from within South Vietnam—not externally from North Vietnam.[44] Therefore, even if the VC had been cut off from their support in the North, a dubious proposition at best given the extensive and porous South Vietnamese border, it still would have only interdicted less than half of the men and material the VC relied upon for sustained operations.

After 1968, though, what might have been correctly described as insurgent sanctuaries previously were transformed into invasion-staging bases. These invasion-staging bases were not primarily for South Vietnamese insurgent bands hiding from the state. These bases served North Vietnamese conventional combat formations to conduct invasions in

1972 and 1975. This becomes more a discussion of state-to-state warfare between conventional combat formations and less a discussion of insurgency and COIN. The cross-border sanctuaries used by the NVA in 1972 and 1975 are less analogous to the Sandinistas in Coast Rica or Guatemala from 1927 to 1933 and more analogous to the German invasion of Poland in World War II.

The US dealt with the NVA invasion staging bases by conducting spoiling attacks into Laos and Cambodia in 1970 and 1971.[45] However, after 1973, the South Vietnamese did little to interdict these same staging bases. Therefore, the problematic issue with the insurgent sanctuaries and North Vietnamese staging bases was not primarily that the South Vietnamese were not allowed to interdict them. The problem was that the US had developed an RVNAF that was unable or unwilling to deal with them. Certainly, there had been a restriction on US forces from pursuing North Vietnamese formations and South Vietnamese insurgents across the border into Laos, Cambodia, and North Vietnam. However, once the US left, there was no such restriction on the RVNAF.

Continued Support After Withdrawal

The suspension of US combat support and combat-service support to the Government of South Vietnam does help explain the anemic SLAW of South Vietnam. South Vietnam faced two major North Vietnamese invasions after the last US combat formations were withdrawn from combat and or the country. The RVNAF were able to defeat the first NVA invasion in 1972. They were unable to defeat the second invasion in 1975. The principle difference between the relative success of 1972 and abject failure of 1975 was the cessation of US combat support and combat-service support.

Prior to the overrunning of South Vietnam by North Vietnam in 1975, the US had already provided economic assistance in FY1974 to the tune of $6.03 billion (all dollar amounts are in 2019 dollars).[46] The US government also debated the provision of further economic aid in FY1975. Communist aid in 1974 had decreased from a high of $4.706 billion in 1968 to around $1.244 billion per year from 1970–1971.[47] The CIA, Department of State, and Department of the Treasury questioned what would be required per year to ensure South Vietnamese viability. The range these agencies promulgated anticipated that it would require US spending between $2.34–$3.28 billion per year to keep South Vietnam afloat.[48] However, the question of how much it would cost to keep South Vietnam afloat became moot as Saigon and the South Vietnamese government fell five months before the start of FY1975.

The suspension of strategic rents in the forms of combat support and combat-service support had the most injurious effects on the ability of the South Vietnamese to defend themselves. US economic aid was not instrumental in the failure of the South Vietnamese government. In their 1974 assessment of the situation in Southeast Asia, CIA analysts offered that:

> Current supply stockpiles are more than adequate for current consumption rates. Should major countrywide fighting resume, however, the South Vietnamese logistic system would be stretched to the limit...ARVN remains dependent on US civilian contractors for aircraft maintenance and port management... Serious problems still exist, however, especially those stemming from inexperience in managing, supplying, and maintaining a large force in combat.[49]

The "serious problems" ... "stemming from inexperience" logistically sustaining "a large force in combat" should have been inconsistent with the history of the war to that point. By the time of this assessment, the RVNAF had been trained by the US in logistics, supply, and maintenance for two decades. Moreover, at its max limit, the RVNAF only had to transport its logistics using interior lines across decent roads a maximum of around 400 miles. In contrast, the North Vietnamese had to do the same through jungles, mountains, and trails for 700–1000 miles while under constant US interdiction. However, the CIA assessment points to one key element for South Vietnamese failure—logistical ineptitude.

Based on comparing the performances of the RVNAF in 1972 and in 1975, there is sufficient evidence to conclude that the South Vietnamese government could have defeated the North Vietnamese offensive in 1975 —but only provided they had received the same degree of combat support as in 1972. In addition to logistical gaps, the withdrawal of US forces left a massive gap in aviation sorties just to maintain the level of support that kept the RVNAF alive in 1972. The withdrawal of the US aviation support left an air support gap of 423,000 armed helicopter sorties, 2,045,000 combat support sorties, 1,744,000 noncombat support sorties, and 121,000 in tactical air sorties from 1972–1975.[50] Indeed, it could be reasonably argued that the South Vietnamese might have persisted indefinitely if the US had been willing to provide this massive combat support indefinitely.

In contrast, the North Vietnamese government and military received a comparatively microscopic amount of economic and military aid from the Soviet Union and the People's Republic of China. Additionally, the USSR and the People's Republic of China provided no combat formations and no frontline advisors in diametric contrast to the massive interventions by US combat formations. The CIA estimated that at its highpoint, the government of North Vietnam received a maximum of $4.076 billion in economic and military aid. The aid originated from the USSR and the People's Republic of China in 1967–1968.[51] From 1971 on, the CIA estimated that the North Vietnamese government received around $1.244 billion for the duration of the war from the same sources.[52] North Vietnam also received a relative handful of Soviets trainers/advisors. The CIA estimated in 1965 that between 1,000–1,500 Soviet personnel were initially operating the surface-to-air missile systems in North Vietnam.[53] However, the Soviet advisors appear to have rapidly turned these systems over to North Vietnamese troops. Therefore, despite an exponentially greater investment in money and personnel, the US was nevertheless unable to create a South Vietnamese government capable of defending itself against an adversary receiving the tiniest fraction of similar support.

GOVERNANCE AND SECURITY FORCE DEVELOPMENT

Degrees of Embeddedness in Governance Development
From 1954 to 1973, the US military and US Department of State relied on institution-influencing relationships to develop the Government of South Vietnam. These low embedded strategies included coercive influence and advise-and-assist relationships to increase South Vietnamese government's capacity. Whereas the US assumed full control of the governance of Germany, Austria, Italy, and Japan to develop governance in years after World War II, the US eschewed this in Vietnam. The choice in US methods to develop South Vietnamese governance and security were reactions to Marxist critiques both in Vietnam and globally. Lieutenant Colonel Robert M. Montague Jr. specifically captures the zeitgeist concerning socialist critiques of colonialism and imperialism in Vietnam by asserting that "communist propaganda was playing the theme that Khanh, like Diem before him, was a US 'lackey,' installing district advisors would surely give Viet Cong propaganda new neo-colonial teeth."[54] US leaders sought to at least preserve the façade of South Vietnamese sovereignty in light of these Communist critiques. Yet American efforts to rehabilitate South Vietnamese governance resulted in what I term a sovereignty fallacy. Further, after two decades of US efforts to advise and influence South Vietnamese institutions from the outside-in, these institutions were no more capable of governing and defending South Vietnam in 1975 than they were in 1954.

From 1954 to 1973, the United States employed a combination of coercive influence and advice and assistance to develop South Vietnam's government. President Diem's and the Government of South Vietnam's ideological background was both anti-colonial and anti-communist. This made the South Vietnamese particularly sensitive to anything that approximated colonial rule by the US or France.[55] Whereas the South Vietnamese government had no inhibitions accepting billions of US dollars in aid and hundreds of thousands of US troops in assisting them in killing rebellious South Vietnamese on their behalf, these same leaders often blanched under US advice. Specifically, Montague asserts that "to some Vietnamese leaders, advice was anathema—something that had to be tolerated in exchange for the huge quantities of US aid which their country was receiving."[56]

Therefore, in order to obligate the Government of South Vietnam to accept US advice, US diplomats and civilian and military advisors relied on coercive influence rather than direct control. John Mecklin, former Director of the US Information Services in Saigon, related "that economic and military assistance was a strategic government-to-government operation, especially in a struggle against guerrillas. Aid should be given only on the specific understanding that American advice on its use must be heeded all the way down to the point where the last cartload of fertilizer was delivered to the peasant."[57] Mecklin advised that receipt of US aid was and should be contingent on South Vietnamese acceptance of US advice.

As a result of South Vietnamese reluctance to accept US influence, a see-saw relationship ensued between the US advisors and senior South Vietnamese leaders. Even as President Diem often refused US advice, the CIA noted that Diem was intensely dependent upon US economic and military aid and could not afford to allow this aid to be interrupted. However, the CIA also noted that Diem was also well aware of where his place and South Vietnam's stood in American foreign policy. As a result, a precarious balance existed where both the US and the South Vietnamese government were intrinsically reliant on each other but incapable of asserting complete control.

The US government wanted both control over the South Vietnamese government and the veneer of South Vietnamese sovereignty. This produced what I describe as a sovereignty fallacy. The most important

The Crumbling State 185

sovereignty contradiction is also the most obvious. At the high point of the war, over 536,000 American troops operated largely unilaterally within South Vietnam, liquidating thousands of rebellious Vietnamese on behalf of the South Vietnamese government. It should have been difficult for the Government of South Vietnam to claim any reasonable pretense of sovereignty while the assistance of US combat formations was required to kill rebellious Vietnamese citizens (insurgents) on its behalf.

Another less obvious dissonance between US control and South Vietnamese sovereignty came about as a result of the discussions regarding the structuring of the RVNAF. The Government of South Vietnam preferred territorially associated units along the French colonial model. The US preferred strategically mobile, national forces.[58] The US advisors saw attack by the NVA as a greater threat and preferred a large conventional force that would be able to defeat North Vietnamese invasion.[59] Because the US was paying the bills in April 1955, the decision was made for more mobile national forces.

Another example of sovereignty fallacy concerned who had the authority to increase the size of the RVNAF. It was President Kennedy, and not South Vietnamese President Diem, who approved an increase in the ARVN from 150,000 to 170,000 soldiers in April 1961.[60] Then, even more implausibly, when Diem wanted to increase the size of his own ARVN to 270,000 troops, he had to request permission from President Kennedy to do so. Kennedy did not approve Diem's request.[61] In essence, a US president refused permission for a foreign president to increase the size of his own forces.

Finally, President Nixon, not the South Vietnamese government, directed the ARVN raid into Laos in February 1971. This is particularly instructive because the operation relied exclusively on Vietnamese ground forces (in Laos) but was directed from Washington DC, not Saigon.[62] No US ground forces were allowed on the mission into Laos due to the US congressional injunctions of January 1971[63] prohibiting the employment of US troops outside South Vietnam. However, this did not prevent the

US president from ordering the forces of another state to enter another foreign state to fight other foreign forces. The US president, not the Vietnamese president, directed an exclusively Vietnamese operation. It is clear that even though the US eschewed the use of high embedded strategies to develop the South Vietnamese government, the South Vietnamese sovereignty was only an affectation. This calls into question the utility of this pretense of sovereignty in developing and advising the South Vietnamese government.

US military and civilian officers advised and assisted the South Vietnamese government at the national and provincial levels from 1967 to 1973 through a program called Civil Operations and Revolutionary Development Support (CORDS). Whereas CORDS has been well regarded by many observers,[64] others see its successes as too little, too late[65] or unable to overcome the systemic weaknesses in the South Vietnamese system.[66] Frank L. Jones of the US Army War College argues that the insufficiency in CORDS came not from its implementation or strategy, but from the system it was introduced into. He argues that:

> Despite these positive outcomes... Pacification could not 'cause a fundamental transformation of South Vietnam,' and the ultimate goal of pacification was to transform the government structure into a system that could attain popular support. The US-backed pacification program could not overcome the South Vietnamese government's defective execution of plans and programs, its omnipresent corruption, or its inability to develop a sturdy, self-sustaining political base.[67]

Still, other than the Marine Corps' Combined Action Platoons, CORDS remains one of the few programs to receive positive affirmation in contemporary analysis of the development of the South Vietnamese government.

The union of the US military's access, security, and resources with the US government's civilian subject matter expertise allowed CORDS to make material progress in 1967. It had been deficient since the US involvement began in 1954. A singular attribute of CORDS was its thorough integration of US civilian and military advisors into a single unified organization. Notwithstanding the disproportionate ratio of military to civilian personnel (6,500 military to 1,000 civilian), CORDS represented an unprecedented fusion of civil-military operations. The program was led by a civilian, Robert Komer, and his stature within the military's structure was also unprecedented. Komer, though a civilian, was appointed as one of General Westmoreland's three deputies, accorded the title of ambassador, and vouchsafed a rank equivalent to a three-star general.[68] Komer recognized that the civilian experts in governance, rule of law, economy, and infrastructure development possessed superior capacity in these areas compared to their military peers. However, these civilian experts lacked the security, access, and resources that only the US military possessed in Vietnam.[69] This union of civil-military operations was trained and forged at the Vietnam Training Center (VTC) located within the Foreign Service Institute in Arlington, Virginia, in the 1960s.[70]

Though the CORDS program yielded tangible results in a whole-of-government approach to pacification, it was constrained to invest a substantial amount of time and effort toward the paramount issue of security. According to John Paul Vann, the Senior CORDS administrator, "Whether security is ten percent of the total problem or ninety percent, it is inescapably the first percent of the first ninety percent."[71] The core element in CORDS' security line of effort was the territorial or paramilitary forces—the Regional Forces/Popular Forces (RF/PFs). By 1972, the US had trained over 532,000 RF/PFs, and these would constitute over half of the total 1,048,000 strong RVNAF.[72] The RF/PFs would not only contribute disproportionately to the pacification of South Vietnamese villages and hamlets but would also exceed the large conventional forces in their ability to attrite VC/NVA forces.[73] The RF/PFs were trained and advised by a number of different US efforts. The first, beginning in 1965, was the District Advisors Program (DAP). Small elements of two-to-seven-member teams of US soldiers would split their time between advising the RF/PFs and advising the district chief.[74] Another effort was provided by US Army mobile training teams (MTTs). These would superficially develop RF/PFs by staying with them a few weeks before moving on to train the next team.[75]

A more comprehensive method of developing the RF/PFs was embedded or parallel advising efforts under the Combined Action Platoons (CAPs). The RF/PFs were most effective when they were developed by Marines who lived on a daily basis in the villages with the RF/PFs they were training. With only fifteen Marines and thirty-four RF/PFs, the CAPs were able to turn the tables on the VC. The CAP program forced the VC to resort to large-scale unsuccessful operations to dislodge the CAPs from villages.[76] Only one village would be overrun during the entire program, despite the commitment of entire VC battalions to attack mere RF/PF platoons.[77] General Westmoreland eschewed the CAP concept despite the wholehearted approval of no less than the preeminent doyen of COIN of the 1960s, Sir Robert Thompson, who asserted that "the use of CAPs is quite the best idea I have seen in Vietnam, and it worked superbly."[78] As a result of Westmoreland's aversion to the CAPs, the program's successes were limited to only a few villages in the Marines' area of responsibility.

CORDS security efforts would also be advanced through the development of police, the recruiting of VC defectors as part of the Chieu Hoi program, and the destruction of the VCI. Under CORDS, the US advisors would oversee the creation of 39,000 new national police.[79] However, systemic corruption, inadequate funding, and poor recruiting resulted in a force wholly inadequate to compete with the VCI.[80] These police were also disproportionately employed in already pacified areas rather than areas where their efforts were more desperately needed. This led to the most problematic areas being largely ignored.

A far more successful CORDS program was the Chieu Hoi, or Open Arms, insurgent amnesty program. The Chieu Hoi program encouraged the defection of the former VC. From the start of the program under President Diem until 1967, 75,219 Hoi Chanh surrendered. Most of these were former VC combatants rather than mere political supporters.[81]

The Crumbling State   189

These numbers of Hoi Chanh would constitute the equivalent of the surrender of two entire VC battalions in one year alone. The impact of the Hoi Chanh, or the amnestied insurgents, was threefold. First, they attrited VC ranks without creating enemies of the dead VC family members. Second, they crushed morale among the VC. VC defectors generated far more propaganda value by their defection than the killing of the VC did. They did this by undermining VC and peasant belief in the cause. Thirdly, Hoi Chanh provided an intelligence coup. Hoi Chanh could deliver over their former comrades and help to more effectively hunt down other VC.[82]

The Phung Hoang, or Phoenix Program, unlike the national police or RF/PFs, was singularly focused on and successful in destroying the VCI. The VCI consisted of the insurgent shadow government as well as the logistical structure that supported VC combat forces in South Vietnam. Even as the larger CORDS focused on security and development of South Vietnamese governance, Phoenix strove to obliterate the VCI, the Government of South Vietnam's competition. Where ARVN, US forces, and RF/PFs targeted main force VC combatants, Phoenix targeted political cadre members, tax collectors, recruiters, and terrorists. Former *Time* and *Reuters* journalist and North Vietnamese spy, Pham Xuan An, attested to the Phoenix Program's efficacy. He related that as a result of the Phoenix Program, the VCI had been so thoroughly eviscerated in South Vietnam that the North Vietnamese had to surge Communist cadre to the south after the fall of Saigon to govern these areas.[83] According to Andrew Finlayson, who was detailed from the US Marine Corps to the CIA program, two former Vietnamese Phoenix officers who escaped Vietnam in 1982 also corroborated An's observation.

CORDS also pursued agricultural development, humanitarian assistance, infrastructure and economic development, and training of bureaucratic administration.[84] In many studies, the security efforts executed under CORDS receive the lion share of attention. However, CORDS was designed as an "innovative whole-of-government approach to achieving rural pacification through development activities strategically coordinated with military operations."[85] CORDS worked through US agencies like the US Agency for International Development to develop South Vietnamese governance capacities.[86]

US civilians only accounted for 2,685 personnel operating in Vietnam in 1969, and over half of these were associated with CORDS.[87] In comparison to the 475,000-plus American military personnel in Vietnam at the same time, the number of civilian experts pale in comparison. The relative paucity of civilian experts therefore necessarily limited civilian impact on governance development to strictly advising and provision of funds.

CORDS was principally weakened by the pervasive institutional corruption of the South Vietnamese political and military systems. As Honn et al noted, "the political corruption and instability in the South Vietnamese government was a fundamental problem that hampered the success of the CORDS program as well as the US's ultimate goal of pacification. Individual CORDS workers found that many of their Vietnamese counterparts were more interested in their own personal wealth and well-being than in pacification or the implementation of development projects."[88] This sentiment is echoed through the USAID assessment of CORDS as well.[89] Corruption was systemic through every level of civil service. Bernard B. Fall described Communist revolutionary war as essentially political war.[90] From this perspective, the cardinal vulnerability of the South Vietnamese war effort was not tactics or logistics, but corruption.

Whereas initial CORDS efforts worked mostly at the provincial and national levels, the bulk of the pacification strategy was executed at the district-chief level. Initially, there was no US structure in place to advise down to the district level.[91] Lieutenant Colonel Montague, a district advisor from 1965 to 1966, cataloged the disquiet he and other US agents had regarding the situation in Vietnam in 1964.

Montague noted that by 1964 the US recognized that the pacification war was being won or lost at the district level and not at the provincial or national levels. He noted that where the district chiefs were effective,

The Crumbling State 191

provincial governance and security efforts were also proportionally more effective. US advisors had previously only provided military assistance to governance down to the provincial levels until that time.[92]

District Advisory Teams were designed to close the gap between the provincial and village levels. Military personnel would directly advise the district chiefs in their wide array of responsibilities. District Advisory Team members had no unique training or special skill sets. However, unlike their US civilian counterparts, they were able to operate in more dangerous environs. In April 1964, the US instituted a pilot District Advisory Program with thirteen two-man teams made up of one US officer and one sergeant. US district advisors spent three-quarters of their time advising the district chiefs on non-military matters despite possessing no unique qualifications to do so.[93]

The District Advisory Teams produced significant short-term results, but these proved to be unsustainable in the long run. Montague noted that the South Vietnamese government's logistical system that reached down to the districts was feeble, corrupt, and easily exploited. The presence of US officers allowed the South Vietnamese logistical system to be artificially supported by the more robust American military system. Communications between the district and provinces improved principally because US officers at the district level had to maintain continuous radio communications with their US commanders at the provincial level. As a result, district messages could be communicated via the US military system, and US commanders operating at the provincial level would pressure provincial chiefs to act more rapidly on district requests. Additionally, because of the need to move US troops from district to provincial levels on a regular basis, this also facilitated moving district chiefs and their staffs as well. This increased the person-to-person connections between the formerly isolated districts and the provinces.

As a result of the perceived successes of the pilot program in 1964–1965, the District Advisor Program exploded from thirteen teams to more than 140 teams in 1966. The District Advisory Program was a topic of discussion during the 1964 Honolulu Conference between Westmoreland and Secretary of Defense Robert McNamara. During this conference, both men decided to increase the program. In addition to increasing the number of teams deployed, the size of the teams was also increased from two soldiers to six. The value of the relationships developed by the district advisors is attested to by the reliance of the district chiefs on many of their advisors, despite the advisors' lack of any specific civilian expertise.[94]

In spite of the apparent successes of the District Advisor Program, it suffered from two overriding shortcomings: 1) the impermanence of the improvements brought by the advisors, and 2) the pervasive nature of institutional corruption in the South Vietnamese system. The improvement in district and provincial communications was entirely associated with the temporary presence of the advisors. The district advisors did not ameliorate the dysfunctional institutional communication architecture but only provided a temporary fix to an interminable problem. Similarly, US military logistical systems did not repair the broken Vietnamese civilian system. Instead, it further enfeebled the Vietnamese system as bureaucrats avoided using the decrepit Vietnamese system in favor of the robust, but stop-gap, US system. Finally, and most consequentially, the extensive corruption in the South Vietnamese system was only mitigated, to whatever degree it was mitigated, when US officers were present.

Degrees of Embeddedness in Security Development
From 1954 to 1973, the US military influenced and advised the development of the armed forces of South Vietnam. The choice of this strategy was deliberate. On April 15, 1965, President Lyndon Johnson requested investigation into the potential use of encadrement to develop the RVNAF as an option to forestall the need to invest large US combat formations in Vietnam. Yet, it took only three days for General Westmoreland and his deputy, Lieutenant General Thockmorton, to summarily reject the idea. Thockmorton argued that "there would be both language difficulties and increased support requirements for American personnel under such

a plan. He recommended that the experiment not be tried: American soldiers simply did not lead native forces."[95]

Thockmorton's argument is spurious at best. Language problems had hardly stopped American-Vietnamese advising for the eleven years prior to this discussion. Further, language problems would persist no matter what strategy the US pursued. In addition, the force increases of 1965–1966 created far more support requirements than encadrement would have. Finally, the American military in 1965 had over thirty-five years of experience in leading native forces from 1898 to 1933. Therefore, contrary to Thockmorton's reservations, the US would not have borne any additional risk than it had already accepted, and it may have produced a cheaper and more enduring success. Nevertheless, Westmoreland and Thockmorton explicitly disavowed an institution-inhabiting strategy.

The US developed a Vietnamese military that was no more capable of defending itself against its northern neighbor in 1975 than it had been when the US began to develop the RVNAF in 1954. This poor showing was in spite of a National Security Council assessment in 1972 asserting that "in assessing RVNAF strength, the CIA concludes that both in the present period and in early 1973 ARVN should—from a quantitative point of view—be able to handle the internal security demands as well as the main forces threat."[96] At every point in the war, the South Vietnamese possessed a quantitative advantage over the North Vietnamese forces operating in South Vietnam.

In 1954, the US military in Vietnam identified the greatest threat to South Vietnam as a conventional invasion from North Vietnam. The US anticipated an invasion along the same lines as North Korea in 1950. Despite South Vietnamese preference for a French-styled light infantry model, US planners developed an American-styled military designed to defeat a northern attack.[97] The South Vietnamese military that would evolve over twenty years would resemble a westernized conventional military in every way except command and control, logistics, aviation support, and an economic base able to sustain a highly industrialized military. Thus, from 1950 to 1956, the RVNAF would be organized in an ARVN, a Vietnamese Air Force, a Vietnamese Navy, a Vietnamese Marine Corps, as well as Civil Defense Guards and Self-Defense Corps.[98]

Command and control and logistics were two crucial RVNAF shortfalls that US advisors attempted but failed to influence from start to finish. This failure resulted in confused combat reporting from the front, confusion in authorities, confusion in responsibilities, a lack of accountability, and glacial decision-making during crises. From 1954 until 1975, US advisors were unable to compel or convince South Vietnamese military and civilian leaders to ameliorate these shortfalls.

Unlike the US military, unity of command, unity of effort, and delegation of authority to subordinate commanders were not esteemed in the same way by French colonial rulers in Vietnam, President Diem, or the RVNAF. The French colonial system was designed to prevent independence and the consolidation of power by Vietnamese challengers to French rule. This resulted in a byzantine chain of command from the national level through regional power nodes. The system the South Vietnamese inherited provided for direct coordination and control among more than forty provincial chiefs, functional commands, combat commanders, and the president himself. Within this system, a mid-level commander could receive contradictory orders from several commanders at the same time and even directly from the president himself. In April 1961, President Kennedy sent Vice President Johnson to warn that future US economic and military support would be conditioned upon reform of the Vietnamese national command and control system. Diem agreed, but little changed.[99] Further, politics, rather than merit, was a more consequential criterion for promotion and a severe impediment to strengthening the officer corps.[100]

The US failure to develop the Vietnamese military logistics system began in 1955 under the Temporary Equipment Recovery Mission (TERM) and culminated during the NVA invasion of 1975. From the start, US advisors had been assigned to all major South Vietnamese logistics organizations. South Vietnamese officers were also sent to US service schools for supply and maintenance management training. Brigadier James Lawton Collins Jr. noted that while American advisors observed some successes in the Vietnamese medical corps, quartermaster services, ordnance, and military engineers, they did not see the same successes in transportation and signals. Collins related that the transformation of the South Vietnamese logistical system was complete by 1957.[101] In spite of this sanguine assessment, the South Vietnamese logistical system would remain a critical vulnerability for the RVNAF until the very end.

In 1961, General Maxwell B. Taylor would argue that the South Vietnamese logistical system was already failing and needed US support. Again, in 1972 and 1974, South Vietnamese logistics were cited as areas of critical need for US support.[102] Further, the CIA estimated in 1974 that the RVNAF had sufficient ammunition, fuel, and other supplies to fight for an extended period of time but that "at a minimum, large-scale US logistic support would be required..."[103] Massive systemic corruption seems to have been the critical element in the failure of the South Vietnamese logistical system.[104]

The RVNAF were designed, trained, and equipped along American lines and expanded across the board in 1954–1967. This period saw a 700% growth in the Navy, 500% growth in the Marine Corps, 400% growth in the Air Force, and 100% growth in the Army and paramilitary forces (see table 9).[105]

By 1967, the US military had overseen the creation of South Vietnamese artillery, armor, command and staff, infantry, administration and finance, adjutant general, quartermaster, ordnance, medical, transportation, intelligence, logistics management, and military band schools.[106] The bulk of the training and operations of this 643,000-man force was focused on pursuing the large VC and NVA main force units in the less populated hinterland of the country.

Table 9. Growth of the RVNAF, 1954–1967.

YEAR	REPUBLIC OF VIETNAM ARMED FORCE STRENGTH 1954-1967								
	Army	Navy	Air Force	Marines	Total Regular Forces	Regional Forces	Popular Forces	Total Paramilitary Forces	Total Force Structure
1954-55	170,000	2,200	3,500	1,500	177,200	54,000	48,000	102,000	279,200
1959-60	136,000	4,300	4,600	2,000	146,000	49,000	48,000	97,000	243,000
1964	220,000	12,000	11,000	7,000	250,000	96,000	168,000	264,000	514,000
1967	303,000	16,000	16,000	8,000	343,000	151,000	149,000	300,000	643,000

Source. Table derived from *The Development and Training of the South Vietnamese Army 1950–1972* (Annex D), by Brigadier General James Lawton Collins, Jr., 1974, Washington, DC: Department of the Army.

The Crumbling State 197

Immediately following the Tet Offensive in 1968, there began a de facto Vietnamization of the formerly Americanized war. From April 1968 on, Secretary of Defense Clark Clifford directed the US military to begin to Vietnamize the war, or to gradually transfer responsibility for operations against the NVA and VC from American to South Vietnamese forces. Tactically, Vietnamization of the war was executed through the US Improvement and Modernization (I&M) plan for the RVNAF. This plan was composed of three phases and based on two assumptions. The first assumption was that the North Vietnamese threat that existed in 1969 would persist. The second assumption was that the US would perpetually support the Government of South Vietnam. The first phase of this transition was approved on October 23, 1968, and the second phase was approved on December 18, 1968.[107] US troop levels also began to drop rapidly from 1968 to 1973. The last US combat formations departed in 1972, and all US military personnel were withdrawn by 1973.[108]

Originally, there were two phases to the I&M plan. Phase I accelerated the quantitative increase in manpower in the RVNAF maneuver battalions. Phase II transitioned US gear to South Vietnamese forces. This equipment included aviation, armor, and artillery systems. Phase II also increased the Vietnamese role in aviation, armor, and artillery operations. After it became clear in 1970–1971 that the US would be withdrawing faster than initially anticipated, a Phase III was added to enhance the ability of the RVNAF to provide its own combat support.[109] However, even as the RVNAF was experiencing its greatest growth, the number of US advisors was also dwindling at the exact same time.[110] By 1970, the two US Marine Divisions remaining in Vietnam redeployed. By 1972, only two brigades of the US Army remained,[111] and these departed Vietnam on August 11, 1972.

There were two phases to Vietnamization. The first was to upgrade the RVNAF with modern American combat systems and firepower such as tanks, artillery, and aircraft. The second phase would be training and advising the RVNAF in the application of these assets against the NVA and VC. As US forces were reduced in 1968–1972, the RVNAF ranks swelled with the largest growth coming from non-ARVN combat support and territorial paramilitary forces. The RVNAF went from 643,000 troops in 1967 to over a million by 1972 (see table 9). In sheer numbers, the territorial paramilitary forces had the largest growth of 232,000 troops. The Republic of Vietnam Air Force tripled in manpower while the Navy grew two and a half times larger. The ARVN grew 107,000 troops but also suffered from widespread desertions. Lastly, the Republic of Vietnam Marine Corps grew 75 percent to 14,000.

A 1972 CIA report titled Net Assessment of North Vietnamese and South Vietnamese Military Forces compared the quantitative capabilities between the Communist forces and RVNAF. The report found that "in terms of equipment and training, as well as in terms of numbers, the South Vietnamese ground forces must be rated as equal or superior to the North Vietnamese forces they are fighting."[112] CIA analysts focused on the 320,000 troops in South Vietnamese maneuver battalions of the ARVN and Vietnamese Marine Corps. They compared these to the 154,000 VC/NVA maneuver forces known to be operating in South Vietnam which provided a 2.1:1 advantage for South Vietnamese forces.[113]

Despite their quantitative advantage, analysts also identified a key qualitative shortfall in the RVNAF that undercut the RVNAF's quantitative advantages. The analysts noted that the ARVN battalions had not been used for major kinetic conventional operations against VC/NVA main force units for the most part since American combat formations arrived. South Vietnamese forces were shielded from defeat or destruction by NVA/VC main force units. US units would battle the NVA/VC main force units, and the ARVN would conduct pacification operations within a perimeter create by US combat formations. This shielding kept South Vietnamese forces safe from destruction but also from learning difficult lessons firsthand.

Table 10. Growth of the RVNAF, 1954–1972.

REPUBLIC OF VIETNAM ARMED FORCE STRENGTH 1968-1972								
YEAR	Army	Navy	Air Force	Marines	Total Regular Forces	Regional Forces	Popular Forces	Total Paramilitary Forces	Total Force
1954-55	170,000	2,200	3,500	1,500	177,200	54,000	48,000	102,000	279,200
1959-60	136,000	4,300	4,600	2,000	146,000	49,000	48,000	97,000	243,000
1964	220,000	12,000	11,000	7,000	250,000	96,000	168,000	264,000	514,000
1967	303,000	16,000	16,000	8,000	343,000	151,000	149,000	300,000	643,000
1968	380,000	19,000	19,000	9,000	427,000	220,000	173,000	393,000	820,000
1969	416,000	30,000	36,000	11,000	493,000	190,000	214,000	404,000	897,000
1970	416,000	40,000	46,000	13,000	515,000	207,000	246,000	453,000	968,000
1971-72	410,000	42,000	50,000	14,000	516,000	284,000	248,000	532,000	1,048,000

Source. Table derived from *The Development and Training of the South Vietnamese Army 1950–1972* (Annex D), by Brigadier General James Lawton Collins, Jr., 1974, Washington, DC: Department of the Army.

In terms of major combat systems, the South Vietnamese government had a solid quantitative advantage in tanks, artillery pieces, armor, and aircraft. The South Vietnamese government possessed an advantage in light artillery (105mm) and medium artillery (155mm) but were at a disadvantage in heavy artillery (175mm). Importantly, by September 1971, the RVNAF possessed almost the same number of artillery battalions as the US had employed in South Vietnam at the peak of the US intervention in December 1968. The RVNAF possessed 216 armored personnel carriers in 1972 and received fifty-four medium tanks in September 1972. CIA analysts were concerned with the difference between South Vietnamese and North Vietnamese heavy artillery but noted that this shortfall would be covered by the provision of South Vietnamese and US tactical aviation sorties.[114]

CIA analysts conducting a comparative analysis between North Vietnamese and South Vietnamese forces noted three cardinal shortfalls in the RVNAF. The first was leadership. The second was the RVNAF's inability to cover the gap between American tactical air support and assault support. And the third was the RVNAF's inability to logistically sustain itself without perpetual US support. With regard to the RVNAF promotion system in 1972, the CIA analysts noted it was gradually becoming more meritocratic. However, by and large, the RVNAF and the South Vietnamese government still disproportionately promoted less competent and less aggressive officers who were more politically connected.

In fact, in trying to bring about the 1971 ARVN Operation Lam Son into Laos, US advisors had little luck in getting senior South Vietnamese officers to work together. Because politically connected senior South Vietnamese officers were put in charge, senior officers of the ARVN and Vietnamese Marine Corps actually delegated command of their own forces to junior officers on the front lines. This absenteeism of frontline senior leadership did not presage success. Lam Son was a failure that demonstrated a principle qualitative deficiency of the RVNAF and Government of South Vietnam: senior leadership. Even as CIA analysts noted that corruption also existed among the North Vietnamese officer corps, it was apparently less rampant.[115]

As part of the 1972 assessment, CIA analysts noted that, in 1971, the US was leaving South Vietnam with a massive gap in strategic bombing, close air support, and combat support sorties (as previously mentioned). The RVNAF provided a miniscule portion of the total sorties in each category, and the degree of effectiveness in each was largely subpar. In 1972, the gap in these sorties would be closed by American sorties originating from off-shore bases. But in 1975, the RVNAF would have to close this gap by itself. The CIA's assessment in 1972 specifically observed that even though the Vietnamese Air force "has become steadily more effective...A high level of support is still required from the US government and US contractors...Furthermore, neither now nor in 1973 will the VNAF be capable of providing sufficient air support."[116]

Lastly, whereas both North and South Vietnam received foreign aid, the South Vietnamese government received substantially more and depended on it more. North Vietnam received significant technical and economic/military support from the USSR and China. However, CIA analysts assessed that, due to North Vietnamese stockpiling, even a reduction of 50 percent of all foreign aid received would not inhibit North Vietnamese protracted warfare throughout Indochina.

At the tactical level, CIA analysts noted several critical shortcomings in the South Vietnamese logistical system compared to North Vietnam's. The South Vietnamese logistical system suffered from system-wide corruption and lacked the ability to conduct high-end maintenance on the major combat systems received from the US. In addition, the RVNAF also lacked the ability to repair communications and sophisticated electronic systems provided by the US. Supply management and inventory control were also feeble, and almost half of the water borne transportation of material still had to be moved by US forces in 1972.

Three injurious elements hamstrung US efforts to create a three-dimensional replica of itself: lack of senior leadership development, shielding of RVNAF forces from the most challenging combat and support requirements, and failure to compel the South Vietnamese government to enforce its own desertion and conscription laws. With respect to leadership in the South Vietnamese military, Krepinevich observes that "most conspicuous, however, was the absence of that prime ingredient of successful armies—capable leadership."[117] French colonial rule prohibited Vietnamese officers from rising above the rank of junior officers from the nineteenth century until 1951, when the first formal Vietnamese officer commissioning programs were begun. By 1952, the RVNAF had its first command and general staff course for field-grade officers. The Vietnamese military academy was established in 1956 for regular officers, and reserve officers were trained at the Thu Duc Officer Candidate School. Despite all this, selection of officers was disproportionately focused on political connections and civilian education rather than an officer candidate's ability to lead. Though the RVNAF made some strides in commissioning prior enlisted NCOs as well as candidates with less-aristocratic educations, this was not wide-spread. Additionally, officers in the RVNAF tended to remain in the same units and same pay-grade levels with little rotation between combat and support roles.[118]

The US stunted the institutional learning and growth of the RVNAF after shielding the ARVN from the North Vietnamese forces for over a decade. The RVNAF would later have to face these forces on their own and without US support. From 1965–1971, US forces insulated the anemic RVNAF from the most dangerous fighting. According to General Andrew Goodpaster, "ARVN and RF/PF secured areas we cleared. Our job was to keep the NVA Main Force away from secured areas we cleared; inside them, ARVN, RF/PF, and CORDS."[119] Goodpaster noted that after Tet, the US forces were protecting the ARVN from the NVA main force units. This shielding compelled US and NVA forces to continually adapt and learn in order to get ahead of each other. In contrast, this shielding retarded the RVNAF from having to learn the painful, but necessary, lessons the NVA were forced to learn.

Promises of perpetual US support seem to have prevented the South Vietnamese government from taking a more dramatic stance against its catastrophic manpower shortages. The South Vietnamese government resisted the aggressive enforcement of its already-weak conscription laws despite US entreaties. This resulted in rampant draft evasion and a force hollower than what the North Vietnamese produced through stricter conscription laws. From the start of conscription laws in September 1957 and until 1965, some 232,000 Vietnamese are estimated to have evaded conscription. Additionally, weak desertion laws and flaccid enforcement exacerbated the already challenging RVNAF manpower problem. In 1965 alone, 113,000 Vietnamese deserted the RVNAF. Whereas the rate of desertions would drop by 30% in 1967 and continue to drop, nearly 80,000 Vietnamese would continue to desert the RVNAF every year throughout the remainder of the war.[120] If the numbers of South Vietnamese deserters and conscription evaders are combined from 1954–1974, as many as 1,993,000 South Vietnamese men might have avoided service out of a country of less than 18,000,000.[121] To provide context, in the thirteen years of draft and service in Vietnam (1960–1973) the US had approximately 569,517 draft evaders.[122] That is, the US had only one-third the number of draft dodgers as South Vietnam, despite having a population almost twelve times larger and fighting on foreign soil.

ADVISOR SURVEYS

From December 2016 to May 2017, fifty-five US advisors to the RVNAF were surveyed as part of my research. These advisors served all over South Vietnam, Cambodia, and Laos; and their experiences cover the entire range of the US intervention in South Vietnam, from 1960 until the Easter Offensive of 1972. The advisors were almost entirely commissioned officers, with a small number of senior NCOs, and all served in either the US Army or US Marine Corps. The lowest-ranking advisor surveyed was a senior non-commissioned officer, and the highest-ranking officer interviewed or surveyed was a four-star general.

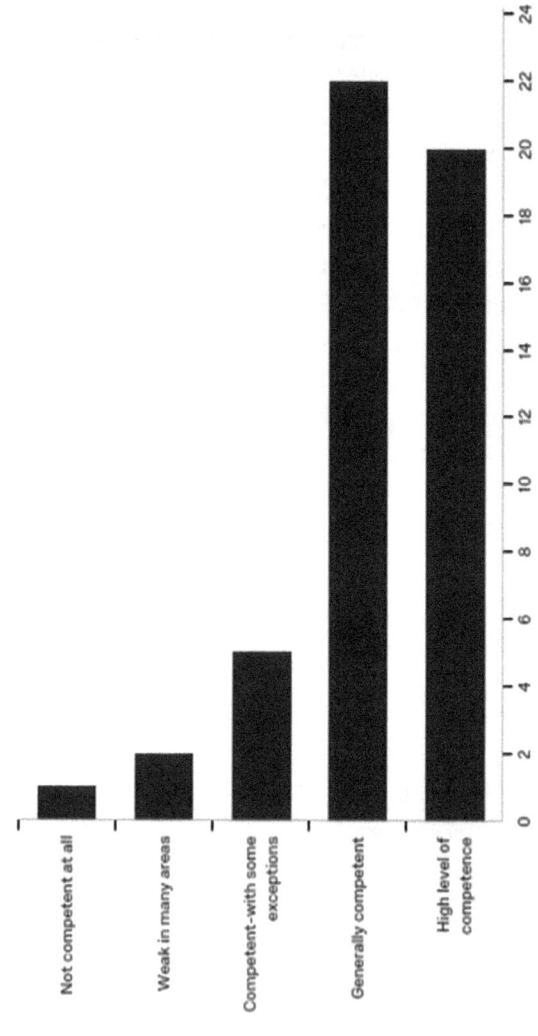

Figure 8. Advisors' Perception of Vietnamese Competence Upon Withdrawal.

Given the poor performance of the RVNAF in 1975, I expected universally negative opinions of the RVNAF by their US Advisors. This assumption was proven dramatically erroneous. Specifically, General Anthony Zinni, then a Marine Lieutenant advising the Vietnamese Marine Corps, related that:

> I was advising the Vietnamese Marine Corps (VNMC), and even after over four decades of military service, I would consider them one of the best light infantry organizations in the world. The company commander I advised was 55 years old, had been wounded nine times, and had numerous awards for valor. The caliber of VMC officers was "off the page." VNMC officers were highly selected, tough, and courageous.

General Zinni's opinion was not an isolated one. In regard to the more elite branches of the RVNAF, advisors across the board spoke highly of them. Additionally, advisors, in general, spoke well of the units they advised, whether or not they advised elite units or conventional elements.

Impressively, when asked about Vietnamese military competence upon the end of the advisors' tours, 84% said the Vietnamese were generally competent to highly competent (see figure 8).[123] Highest praise for South Vietnamese forces was reserved for Vietnamese Marines, members of the Provincial Reconnaissance Units (PRUs) involved in the Phoenix Program, and Vietnamese Rangers. The opinions regarding non-elite RVNAF forces were more muted and aligned well with General Zinni's assessment of them:

> However, this level of quality officer was not uniform throughout the South Vietnamese military. The elites: VNMC, Vietnamese Airborne and Rangers were very good along with other non-elite standouts. However, these exceptional units were the exception rather than the rule. The majority of South Vietnamese forces, in my opinion, were not this high caliber. However, the majority of the South Vietnamese forces could compete effectively with their North Vietnamese and Viet Cong enemies IF there were

properly supported by the US in the American manner they were trained to fight in.

When advisors were asked if the average Vietnamese soldier they advised compared favorably to US soldiers, only ten advisors, or 37%, answered unequivocally no. In contrast, 63% said the Vietnamese troops they advised were as good as US forces, and several said they were better in some ways. This was heavily influenced, however, by US Marine advisors who served with the Vietnamese Marine Corps. They accounted for twelve of the seventeen positive responses. However, when former advisors like General Zinni addressed the more common (non-elite) of the RVNAF, he and the other advisors agreed that, even though these non-elite forces were not the caliber of American forces, they were the equals of the North Vietnamese—if they were supported perpetually by the US. Lastly, when asked about the comparative qualities of the Vietnamese leaders compared to the American leaders, only six, or 30%, of twenty advisors who answered unequivocally said Vietnamese leaders were less capable than their US counterparts. In contrast, 70% said the Vietnamese leaders were unequivocally as good. The advisors found fault most often not at the tactical, or small-unit, level, but at the operational, or strategic, levels of the RVNAF. The most cited concern with the senior Vietnamese leaders was corruption and failure to promote based on merit rather than politics.

When the advisors were asked to compare the quality of their South Vietnamese counterparts to the troops and leaders of the VC and NVA, the majority said their counterparts were better. When asked specifically who had the better soldiers, the insurgents or the South Vietnamese the advisors worked with, of the forty-three who answered unequivocally, 49% rated their counterparts better. Of those who rated their South Vietnamese counterparts unequivocally better, fourteen of these had worked with the Vietnamese Marine Corps, and seven worked with ARVN or RF/PFs. Further, 30% said RVNAF troops were comparable to the VC/NVA, and 23% said the VC/NVA were better. Of those who

The Crumbling State 207

observed that the VC/NVA leadership was better than the RVNAF, most of these specifically identified the senior leadership of the VC/NVA as the reason for their estimate.

Another revelation was advisor perceptions regarding the ability of Vietnamese officers to internalize US leadership traits, principles, and warfare concepts.[124] When asked about difficulty in getting Vietnamese officers to internalize US leadership traits and principles, 74% said it was either possible with effort, possible with minimal effort, or easy. Only 26% said it was impossible or extremely difficult (see figure 9). In breaking out specific leadership traits and principles, over 80% of advisors said getting Vietnamese officers to internalize courage, dependability, and endurance was easy or possible with minimal effort. Similarly, Vietnamese officer judgment, tactical and technical proficiency, and setting the example were also highly rated.

Figure 9. Vietnamese Internalization of US Leadership Traits & Principles.

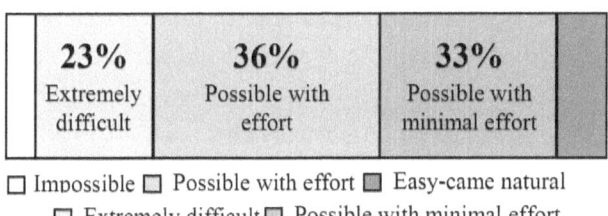

☐ Impossible ☐ Possible with effort ■ Easy-came natural
☐ Extremely difficult ■ Possible with minimal effort

Likewise, when asked how difficult it was to get Vietnamese officers to internalize American tactical concepts, 78% recalled this as being possible with minimal effort or easy. Another 13% said it was possible with effort (see figure 10). Further, when asked to assess how well their Vietnamese counterparts internalized US tactical disciplines, 70% said it was possible with minimal effort or easy (see figure 11). Where advisors took issue with Vietnamese ability or willingness to apply US tactical disciplines, they said it was extremely difficult or impossible to get the Vietnamese to perform advanced maintenance, inspections, and training.

If the advisors took issue with the South Vietnamese officers they advised at all, it was most often in the area of corruption. Whereas 53% of the advisors said the officers they worked with were not corrupt at all, 41% said they were more corrupt than what would be tolerable in the US military. Only 6% indicated their counterparts were criminally corrupt (see figure 12). However, when specifically asked about their dealings with the government and the population, over half (51% and 52% respectively) said their counterparts were either more corrupt than what the US military would tolerate or criminally corrupt even by Vietnamese standards. When asked what types of corruption they witnessed, advisors most often cited the siphoning from subordinates' pay, procuring food from local citizens, receiving pay for dead soldiers, and misappropriation of government material and funds.

Lastly, advisors were deliberately asked an open-ended question regarding what one thing could have been done better regarding the US involvement in Vietnam. This open-ended narrative question was designed to identify if there was a common theme amongst the opinion of the advisors—there was. Twenty-one of the thirty-three who answered the question cited the US's failure to keep its promise of perpetual combat support to the South Vietnamese government as the single greatest cause of long-term failure.

Ascribing the failure of the South Vietnamese government in 1975 to the US cessation of perpetual combat support seems consistent with Krepinevich's assertion. He argues that with regard to the RVNAF, the final product was a westernized military, except that "it did not have the heavy firepower associated with the field armies of Western Europe and the United States..."[125] This also seems consistent with the totality of the US advisor's responses that the US had created a near facsimile of itself with several key shortfalls.

Figure 10. US Advisor Perceptions of South Vietnamese Officer Internalization of US Tactical Concepts.

	Impossible	Extremely difficult	Possible with effort	Possible-minimal effort	Easy-it came naturally
Mission Command	0%	7.14%	9.52%	38.10%	45.24%
Maneuver Warfare	0%	7.32%	17.07%	31.71%	43.90%
Combined Arms	0%	14.29%	11.90%	40.48%	33.33%
Fire and Maneuver	0%	7.14%	11.90%	26.19%	54.76%

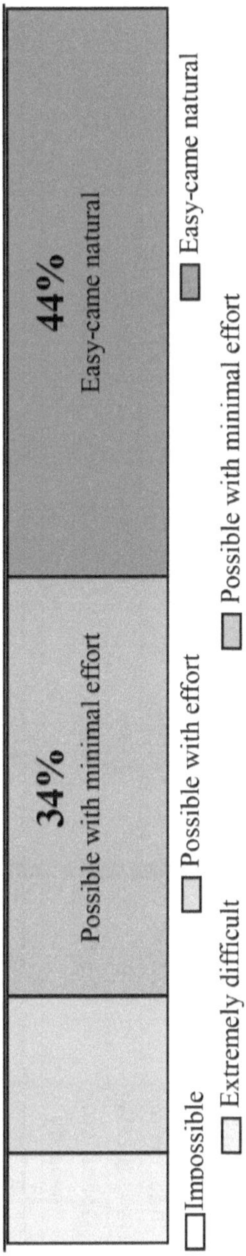

34% Possible with minimal effort

44% Easy-came natural

☐ Impossible   ☐ Possible with effort   ■ Easy-came natural
☐ Extremely difficult   ☐ Possible with minimal effort

Figure 11. US Advisor Perceptions of South Vietnamese Application of Tactical Disciplines.

	Impossible	Extremely difficult	Possible with effort	Possible- minimal effort	Easy-it came naturally
Patrolling	0%	2.38%	7.14%	38.10%	52.38%
Training	0%	14.63%	26.83%	34.15%	24.39%
Inspections	0%	17.50%	37.50%	17.50%	27.50%
Basic maint.	0%	7.14%	16.67%	35.71%	40.48%
Adv. maint.	5.41%	24.32%	21.62%	27.03%	22%
Post & relief	0%	2.56%	7.69%	43.59%	46%
Operational security	2.50%	5%	12.50%	32.50%	48%

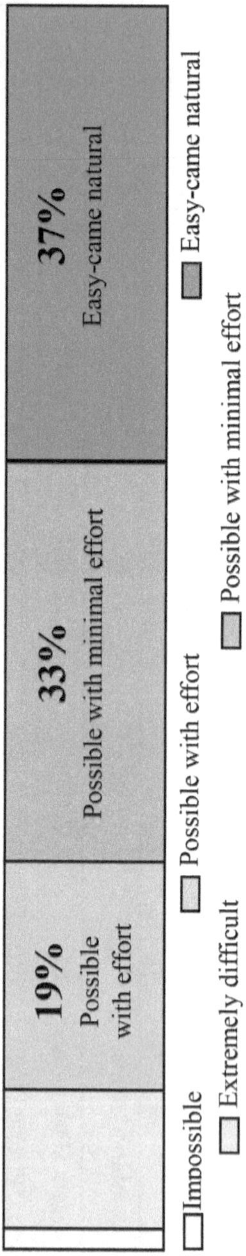

19%	33%	37%
Possible with effort	Possible with minimal effort	Easy-came natural

☐ Impossible ☐ Possible with effort ☐ Easy-came natural
☐ Extremely difficult ☐ Possible with minimal effort

Figure 12. US Advisor Perceptions of South Vietnamese Officer Corruption.

Interactions with:	Not corrupt at all	More corrupt than what is tolerable in the US military, but not criminally corrupt	Criminally corrupt by US military standards
Subordinates	62.22%	33.33%	4.44%
Populace	48.84%	45.51%	4.65%
Government	47.50%	42.50%	10%

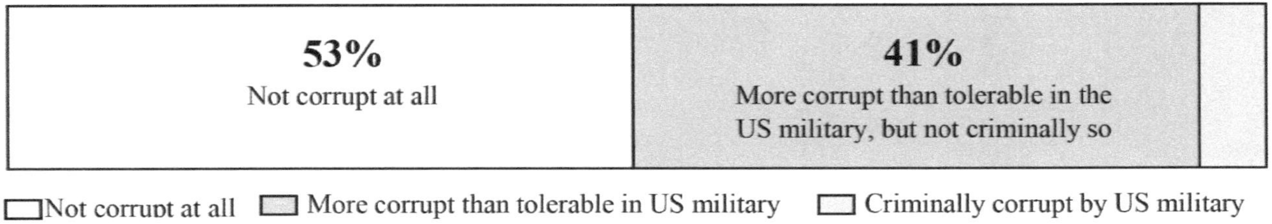

53%
Not corrupt at all

41%
More corrupt than tolerable in the US military, but not criminally so

☐ Not corrupt at all ☐ More corrupt than tolerable in US military ☐ Criminally corrupt by US military

From these advisor surveys, a couple of critical assessments become apparent. The first assessment was that there was a significant departure between what would be tolerated as an American officer and what was common as a Vietnamese officer. This is critical because, as was argued in chapter 3, the American style of warfare relies on a great deal of trust between seniors, subordinates, military officers, and civilian overseers. The American style of war does not appear to translate well into other contexts where corruption is rife and trust is limited. The second assessment that becomes visible is that in attempting to artificially transfer US military and political adaptations, the US created a permanently dependent strategic rentier state in Vietnam. As long as the South Vietnamese government continued to receive strategic rents, it could continue to fight as it was trained to by the US. However, when these rents were withheld, the South Vietnamese government ceased to be able to defend itself.

THE PRE-TEST

From March to October 1972, the RVNAF, supported by 24,000 US advisors, 1,161 US aircraft, six aircraft carriers, and 1,300 South Vietnamese aircraft, defeated a three-pronged, multidivisional NVA invasion supported by tanks and heavy artillery.[126] The deployment of this quantity of US aircraft and naval power, from an initial baseline of only 800 US aircraft in theater, was unprecedented in its scale and timeliness of response. Lieutenant Colonel Matthew C. Brand, US Army Command and General Staff College, argued that the deployment of this much combat power, at the start of the Easter Offensive, constituted "a rapid global mobility response unlike any in the history of warfare up to that time."[127]

In April 1972, the NVA invaded South Vietnam in three phases and in three separate military regions supported by massed artillery, armor, and integrated air defense systems. In all three regions, the ARVN forces either retreated, were severely attrited, or were surrounded and cut off until US strategic and tactical air support broke the back of the NVA

The Crumbling State

attack in each area. In Military Region I, while the VNMC made an orderly retreat south, ARVN forces abandoned Quang Tri province, and the entire 56th ARVN Regiment surrendered en masse. By May, strategic bombing under Operation Linebacker and the US tactical airstrikes punished NVA main force elements and logistical networks.[128]

In Military Region II, the ARVN 22nd Division was savaged by the NVA's second prong, an armor and infantry offensive in the Central Highlands. Only US tactical bombing and intervention by US attack helicopters firing tube-launched, optically tracked, wire guided (TOW) anti-tank missiles turned back the NVA attack, killing twenty-six NVA tanks.[129] The VC/NVA 5th Division, the third prong, attacked central portion of South Vietnam in Military Region III. This attack succeeded in cutting off ARVN forces in An Loc for ninety-five days. This became a pyrrhic victory for the NVA. The successes of these attacks forced NVA main force units to expose themselves to massive US airstrikes. These large-scale conventional combat operations by major North Vietnamese units provided the US something it had lacked for years since the earliest parts of the US intervention: fixed, massed, and easily identified targets. By July, these southernmost NVA forces were decimated and compelled to retreat into Cambodia. From July until October 1972, ARVN forces spent the next few months not only restoring South Vietnamese territory but also seizing portions of North Vietnamese territory.

FIRST REAL TEST—INVASION 1975

After twenty-one years of US training, equipping, and support of the RVNAF, South Vietnamese forces were still no match for the armed forces of North Vietnam without perpetual US support. The Easter Offensive of 1972 provided a test of whether ARVN troops, supported by US logistics and air support, could effectively replace US ground forces in defending South Vietnam. Provided they received US combat support, ARVN could defeat the NVA. However, the NVA invasion that began in December

1974 would test the ability of the RVNAF to entirely replace US ground, aviation, and logistics forces in its own defense.

On December 13, 1974, less than a year after the 1973 Paris Peace Accords went into effect, North Vietnam invaded South Vietnam. The difference between this invasion and the NVA's 1972 invasion would be the lack of US air and logistical support upon which the South Vietnamese government had grown accustomed. The South Vietnamese government was still receiving over $6 billion dollars in US aid annually, in FY1974. The RVNAF were over one million strong. The RVNAF possessed numerical superiority in every metric from infantry, armor, artillery, and aircraft. Initially, the NVA attack consisted of a massive probe designed to test US resolve to intervene. When the US abstained from intervening, the limited attack by the NVA became a full invasion. In less than five months after starting the offensive, the NVA seized Saigon and the South Vietnamese government, and RVNAF would cease to exist.

Conclusion

From 1965 to 1973, the US employed institution-influencing strategies such as advise and assist to develop South Vietnamese security and governance institutions. These strategies of low embeddedness produced host-nation security forces that were only able to act as the senior partner in an oligopoly of violence for less than three years. The result of the US intervention was to produce a strategic rentier state. As long as Vietnam continued to receive strategic rents in the form of close air support, advisors, and massive logistical support, it could and did endure. This was most clearly demonstrated in 1972. However, once the US withdrew these strategic rents beginning in 1973, Vietnam became what I describe as a crumbling state, and by 1975 South Vietnam ceased to exist altogether.

My research finds several observations: 1) the US and the South Vietnamese government did defeat the South Vietnamese insurgency in 1968, relying primarily on enemy-centric COIN methods, and this did little to increase SLAW; 2) the US and the South Vietnamese government did successfully pacify the great majority of South Vietnam in 1968–1972, and this did little to increase SLAW; 3) the US had minimal impact on improving the ability of the South Vietnamese government to govern effectively; 4) in the RVNAF, the US created a two-dimensional facsimile of its own military, but one that lacked the economic and industrial capacity to operate independently; and 5) the US shielded the RVNAF both from destruction and learning, while American combat formations were involved.

The US military and RVNAF decisively defeated the Viet Cong in 1968, and this had little to no impact on the longevity of the South Vietnamese state after US withdrawal. Despite using enemy-centric methods, which have been largely repudiated by the advocates of population-centric COIN theory, the US military and RVNAF confronted and defeated the VC as a result of the Tet Offensive. The US and GVN were also successful in their efforts to pacify the vast majority of South Vietnam. But this also had little to no impact on the durability of the South Vietnamese government after US withdrawal. Moreover, even if the HES quantitative methods have been questioned, it is still clear that pacification was successful in 1968–1972 based on North Vietnam's need to replace VC cadres after the fall of South Vietnam in 1975. Still, even the relative success of pacification does little to explain the inability of the South Vietnamese government to persevere for longer than three years after the US's combat formations withdrew.

The US was not able to materially reform the weak governance of the Government of South Vietnam despite the possession of a significant degree of coercive influence over it. The US failure to improve South Vietnamese governance through influencing relationships did materially circumscribe the longevity of the South Vietnamese state after the withdrawal of US forces. The US was unable to restrain President Diem's repressions of the political and religious dissidents. Diem's heavy-handedness directly led to instability in South Vietnam and materially impacted the government's ability to govern. It also led to the coup that unseated and led to Diem's assassination and compelled the US to intervene with combat formations.

The US was never able to materially address the systemic corruption in the South Vietnamese government or military. Nor was the US able to influence the South Vietnamese government to act for its own self-preservation and strictly enforce its own conscription and desertion laws. Lastly, the US was unable to sufficiently ameliorate the political promotions process of the South Vietnamese government and RVNAF. The South Vietnamese government and RVNAF promotion processes repressed the talented but unconnected while at the same time promoting the incompetent but well connected. This would create a system where US material and economic investments, which dwarfed the investments by communist countries in North Vietnam, were squandered through corruption and mismanagement.

The US created a replica of itself that lacked the economic and industrial resources of the US as well as the military cultural expectations of trust and integrity throughout the US officer corps. These two key shortfalls tangibly reduced the South Vietnamese state longevity after US withdrawal. The United States relies on its industrial and economic base to support combined arms warfare that is incredibly dependent on tactical and strategic air support. The US trained the RVNAF to fight this same style of industrialized and expensive warfare, despite the lack of any similar South Vietnamese industrial or economic base. Moreover, both the RVNAF and American officers were products of the cultures that produced them. The RVNAF officers evolved from French colonial military culture, which devalued delegation of power and authority to subordinates and did not engender trust-based relationships between colonial seniors, subordinates, and adjacent units. In contrast, the entire American style of war is based on the delegation of power and decision making to the lowest levels, in order to generate battlefield tempo.

The Crumbling State 217

Installing an American style of war from the outside-in was difficult, if not impossible.

Finally, the US shielded the RVNAF from both destruction and learning. The NVA was compelled to learn and adapt in its ability to logistically support itself over thousands of miles of trails, roads, rail, and pipeline, while under intense US interdiction efforts. The NVA also had to compete militarily with one of the world's superpowers. Though the NVA rarely, if ever, won any significant battles against US forces, the NVA were nonetheless able to fight the US to a stalemate. After eight years of direct combat against the US military, the NVA benefitted from what Michael Lewis of the Marine Corps University terms a training effect.

During this same eight-year period, the ARVN was only used in mostly pacified areas. The RVNAF was protected from having to figure out how to defeat VC and NVA main force units, how to overcome logistical shortfalls, how to improve aviation combat support, and how to compensate with artillery in anticipation of a reduction in US air support. Shielding the RVNAF from danger and learning resulted in what I delineate a retarding effect. The training effect that the NVA benefitted from, and the retarding effect the RVNAF was hampered by, directly led to what Lewis describes as a training gap and the implosion of the RVNAF in 1975.

The South Vietnamese government was a strategic rentier state from 1954–1973. It was able to defend itself as long as the US continued to supply strategic rents. Further, South Vietnam had become accustomed to receiving greater strategic rents as instability in South Vietnam increased. This created a moral hazard where the South Vietnamese government was actually de-incentivized from improving. This resulted in self-destructive behavior, such as refusing to aggressively enforce conscription and desertion laws. Ultimately, when the US withheld the strategic rents starting in 1973, the Government of South Vietnam became a crumbling state and eventually collapsed with the NVA's 1975 invasion.

Notes

1. Director for Central Intelligence, "Special National Intelligence Estimate 53-64," (1964), https://www.cia.gov/library/readingroom/docs/CIA-RDP79R01012A026000040001-0.pdf, 1.
2. Director for Central Intelligence (1964): 1.
3. Alan Rohn, "How Much Did The Vietnam War Cost?" (April 5, 2016), https://thevietnamwar.info/how-much-vietnam-war-cost/. $28.5 billion in 1974 dollars from 1953–1975, www.in2013dollars.com.
4. Stephen Daggett, *Costs of Major US Wars* (Washington, DC: Congressional Research Service, 2010), 2. $111 billion in 1974 dollars adjusted for inflation covering 1965–1974; www.in2013dollars.com.
5. "Vietnam Statistics - War Costs: Complete Picture Impossible," in CQ Almanac 1975, 31st ed., 301–305 (Washington, DC: Congressional Quarterly, 1976). www.in2013dollars.com.
6. "Vietnam War US Military Fatal Casualty Statistics," (January 11, 2018), https://www.archives.gov/research/military/vietnam-war/casualty-statistics.
7. $650 million in 1974 dollars, www.in2013dollars.com.
8. $450 million in 1974 dollars, www.in2013dollars.com.
9. A. Denny Ellerman, "Memorandum: The South Vietnamese Economy and US Aid. National Security Council (1975)," https;//www.cia.gov/library/readingroom/docs/LOC-HAK-59-2-2-2.pdf.
10. Director of Central Intelligence, "National Intelligence Estimate: The Likelihood of a Major North Vietnamese Offensive Against South Vietnam Before June 30, 1975," (1974), https://www.cia.gov/library/readingroom/docs/DOC 0001166463.pdf, 2.
11. Zinni interview (2017).
12. Congressional Record, "Views of Senator Mansfield on the Crisis in Vietnam," (Washington, DC: US Congressional Record-Senate, 1965), 2557.
13. *Main Force units* is a descriptor unique to Vietnam. *Main Force* units refer to either NVA conventional maneuver battalions *or* Viet Cong conventional maneuver forces.
14. Krepinevich, *The Army and Vietnam*, 56.
15. Ibid., 60–61.

16. Office of National Estimates, "The Consequence of the Attempted Coup in South Vietnam," (November 22, 1960), https://nsarchive2.gwu.edu/NSAEBB/NSAEBB101/index.htm.
17. Office of National Estimates (1960); Prados, J. (2003, November 5). JFK and the Diem Coup. Retrieved from NSA Archives: http://nsarchive2.gwu.edu//NSAEBB/NSAEBB101/index.htm.
18. Collins, *Development and Training of the South Vietnamese Army*, 227.
19. Krepinevich, *The Army and Vietnam*, 197.
20. This is reported by historian Dale Andrade and recorded in *In Persistent Battle: US Marines in Operation Harvest Moon December 8–20, 1965*, edited by Nicholas J. Schlosser (Quantico, VA: Marine Corps, 2017), 4.
21. Krepinevich, *The Army and Vietnam*, 152.
22. Ibid., 151.
23. Schlosser, *In Persistent Battle*, 5.
24. James K. Moore, "North Vietnamese Army's 1972 Eastertide Offensive," (January 9, 2006), http://www.historynet.com/north-vietnamese-armys-1972-eastertide-offensive.htm.
25. Jeremy Patrick White, "Civil Affairs in Vietnam," (January 28, 2009), http://csis.org/files/media/csis/pubs/090130_vietnam_study.pdf, 10–11; Samuel Lipsman and Edward Doyle, *The Vietnam Experience: Fighting for Time* (Boston: Boston Publishing Company, 1984), 74–76.
26. Boot, *Savage Wars of Peace*, 304; Krepinevich, *The Army and Vietnam*, 215, 233; Nagl, *Learning to Eat Soup With A Knife*, 164–166; Lipsman and Doyle, *The Vietnam Experience*, 74–76; White, "Civil Affairs in Vietnam," 10–11.
27. Christopher Fisher, "The Illusion of Progress," *Pacific Historic Review* 75, no. 1 (February 2006): 25–55.
28. Frank L. Jones, "Blowtorch: Robert Komer and the Making of Vietnam Pacification Policy," *Parameters* (Autumn 2005): 103–118; Krepinevich, *The Army and Vietnam*, 233; Nagl, *Learning to Eat Soup With A Knife*, 166.
29. Dale Andrade and James H. Willbanks, "CORDS/Phoenix: Counterinsurgency Lessons From Vietnam For the Future," *Military Review*, 9 (23) (March–April 2006),: 17; *Hamlet Evaluation Survey (HES) Annual Statistical Analysis, 1968–71* (Fort McNair, Washington, DC: Center for Military History, 1968–1971).
30. Andrew R. Finlayson, *Marine Advisors with the Provincial Reconnaissance Units, 1966–1970* (Quantico, VA: History Division, United States Marine Corps, 2009), 52–55.

31. Boot, *Savage Wars of Peace*, 308; Nagl, *Learning to Eat Soup With A Knife*, 167; Krepinevich, *The Army and Vietnam*, 239.
32. Where the VC could no longer viably continue to operate militarily.
33. Andrade and Willbanks, "CORDS/Phoenix," 17; Hamlet Evaluation Survey (HES) Annual Statistical Analysis, 1968–71; Moore, "North Vietnamese Army's 1972 Eastertide Offensive."
34. Mark Moyar, *Phoenix and the Birds of Prey: Counterinsurgency and Counterterrorism in Vietnam* (Nebraska: Bison Books, 2007).
35. Andrew R. Finlayson, "A Retrospective on Counterinsurgency Operations: The Tay Ninh Provincial Reconnaissance Unit and Its Role in the Phoenix Program, 1969–70," (June 12, 2007), https://www.cia.gov/library/center-for-the-study-of-intelligence/csi-publications/csi-studies/studies/vol51no2/a-retrospective-on-counterinsurgency-operations.html.
36. Krepinevich, Nagl, Fisher, and Boot question the capacity of quantitative measures of success for something like COIN that is intrinsically difficult to quantitatively measure.
37. Director of Central Intelligence, "Special National Intelligence Estimate Number 14-69," (16 January 1969), https://www.cia.gov/library/readingroom/docs/CIA-RDP87-00181R000200260009-0.pdf, 5; Krepinevich, *The Army and Vietnam*, 174.
38. Sources: A, B, C—"The Vietnam War: Military Statistics," https://www.gilderlehrman.org/history-by-era/seventies/resources/vietnam-war-military-statistics; *Net Assessment of North Vietnamese and South Vietnamese Military Forces* (April 10, 1972), Washington, DC: Central Intelligence Agency; 1967 CIA estimate of South Vietnamese population: 17,084,771; "South Vietnam Province Population and Area Totals (October 31, 1967)," https://www.cia.gov/library/readingroom/docs/CIA-RDP79-01155A000300020011-3.pdf; D—Size of South Vietnam as of 1967 as estimated by CIA was 65,416 mi2; South Vietnam Province Population and Area Totals, 1967, Washington, DC: Central Intelligence Agency; E—the first number is VC/NVA main force units and the second number is main force units plus support staff; National Intelligence Estimate, North Vietnamese Potential for Fighting in South Vietnam (July 7, 1966), 6, by Director of Central Intelligence (1966), Washington, DC: Central Intelligence Agency; F—CIA memorandum March 18, 1968 estimated VC/NVA total force strength in early 1968 at 500-600k troops in South Vietnam; "Publication of CIA Estimates of VC/NVA Strengths (18 March 1968)," https://www.cia.gov/library/readingroom/docs/CIA-RDP84-00499R000500030003-6.pdf; G, H, I, J—CIA analysis estimated 210 VC/

NVA maneuver battalions were in South Vietnam in early 1968 and then 289 at the end of 1968. It estimated an average of 345 troops per battalion in January 1968 and 265 per battalion in December 1968. This does not account for guerillas or support; "South Vietnam: Impact of Intensified Combat on VC/NVA Maneuver Battalion Strength (August 1969)," https://www.cia.gov/library/readingroom/docs/CIA-RDP85T008 75R001600020111-6.pdf; *Net Assessment of North Vietnamese and South Vietnamese Military Forces* (April 10, 1972), 4, Washington, DC: Central Intelligence Agency; "South Vietnam: A Net Military Assessment, (April 2, 1974)," iii, https://www.cia.gov/library/readingroom/docs/DOC_0001 166747.pdf; "Memorandum for Dr. Kissinger: Net Assessment of North Vietnamese and South Vietnamese Forces," National Security Council (19 April 1972), iv-7), https://www.cia.gov/library/readingroom/docs/ LOC-HAK-451-2-25-1.pdf.
39. Zinni interview (2017); Boot, *Savage Wars of Peace.*
40. Summers, *On Strategy.*
41. Krepinevich, *The Army and Vietnam*, 263; Nagl, *Learning to Eat Soup With A Knife*, 207.
42. Director of Central Intelligence, "National Intelligence Estimate, North Vietnamese Potential for Fighting in South Vietnam," (July 7, 1966), https:// www.cia.gov/library/readingroom/docs/DOC_0000013189.pdf, 10.
43. Central Intelligence Agency, *Daily Supply Requirements of Current and Projected (January 1, 1968) VC/NVA Force in South Vietnam* (August 1, 1966), https://www.cia.gov/library/readingroom/docs/CIA-RDP78S021 49R000200100004-4.pdf.
44. Boot, *Savage Wars of Peace*, 315; Krepinevich, *The Army and Vietnam*, 263.
45. Nagl, *Learning to Eat Soup With A Knife*, 174.
46. Ellerman, "Memorandum: The South Vietnamese Economy," 5.
47. That is $650 million in 1968 and $200 million in 1971. Central Intelligence Agency, *Net Assessment of North Vietnamese and South Vietnamese Military Forces* (April 10, 1972), https://www.cia.gov/library/readingroom/docs/ LOC-HAK-537-7-19-8.pdf, 11.
48. Or $500-700 million in 1975, www.in2013dollars.com. Ellerman, "Memorandum: The South Vietnamese Economy," 6-9.
49. "National Intelligence Estimate" (1974): v–vi.
50. Central Intelligence Agency, *Net Assessment of North Vietnamese and South Vietnamese Military Forces*, VII–7 and VII–10.
51. $650 million in 1968, www.in2013dollars.com. Ibid., 11.

52. $200 million in 1971, www.in2013dollars.com. Ibid.
53. Central Intelligence Agency, "Briefing Notes for the DCI," (October 16, 1965), https://www.cia.gov/library/readingroom/docs/CIA-RDP82R000 25R0006000070007-6.pdf, E-1–E-2.
54. Robert M. Montague Jr., *Advising in Government: An Account of the District Advisory Program in South Vietnam* (Carlisle Barracks, Pennsylvania: US Army War College, 1966), 12.
55. Ibid.
56. Montague, *Advising in Government*, 11.
57. John Mecklin, *Mission in Torment: An Intimate Account of the US Role in Vietnam* (New York: Doubleday, 1965), 314.
58. Krepinevich, *The Army and Vietnam*, 22.
59. Ibid., 23.
60. Ibid., 28 and 56.
61. Ibid., 59.
62. Nagl, *Learning to Eat Soup With A Knife*, 174.
63. *Cooper-Church Amendment* January 1970.
64. Boot, *Savage Wars of Peace*, 304; Krepinevich, *The Army and Vietnam*, 215 and 233; Nagl, *Learning to Eat Soup With A Knife*, 164–166; Lipsman and Doyle, *The Vietnam Experience*, 74–76; White, "Civil Affairs in Vietnam," 10–11.
65. Nagl, *Learning to Eat Soup With A Knife*, xxii.
66. Mandy Honn, Farrah Maisel, Jacleen Mowery, Jennifer Smolin, Minhye Ha, "A Legacy of Vietnam: Lessons from CORDS," *InterAgency Journal*, (2011), 41–50.
67. Jones, "Blowtorch," 116.
68. Andrade and Willbanks, "CORDS/Phoenix," 14; Thomas W. Scoville, "Reorganizing for Pacification Support," United States Army. Center of Military History (January 3, 2006), http://www.history.army.mil/books/Pacification_Spt/Ch5.htm , 63.
69. Nagl, *Learning to Eat Soup With A Knife*, 165.
70. Honn, et al., "A Legacy of Vietnam," 43.
71. Ibid.
72. Collins, *Development and Training of the South Vietnamese Army*, Annex D.
73. Krepinevich, *The Army and Vietnam*, 229.
74. Montague, *Advising in Government*.
75. Krepinevich, *The Army and Vietnam*, 177.

76. Ibid., 174; Bing West, *The Village* (New York: Pocket Books, 2003), 172–173.
77. West, *The Village*, 127–171.
78. Krepinevich, *The Army and Vietnam*, 174.
79. Honn, et al., "A Legacy of Vietnam," 43.
80. Krepnievich, *The Army and Vietnam*, 228.
81. Michael G. Barger, *Psychological Operations Supporting Counterinsurgency: 4th PSYOP Group in Vietnam* (Fort Leavenworth, Kansas: US Army Command and General Staff College, 2007), 38.
82. Finlayson, "Retrospective on Counterinsurgency Operations."
83. Finlayson, *Marine Advisors*, 53.
84. Honn, et al., "A Legacy of Vietnam," 41.
85. Ibid., 49.
86. William P. Schoux, "The Vietnam CORDS Experience: A Model of Successful Civil-Military Partnership?" (September 20, 2005), http://pdf.usaid.gov/pdf_docs/pnaec349.pdf, 15–16.
87. Honn, et al., "A Legacy of Vietnam," 46.
88. Ibid., 48.
89. Schoux, "The Vietnam CORDS Experience," 21.
90. Fall, *Street Without Joy*, 369.
91. Collins, *Development and Training of the South Vietnamese Army*, 49.
92. Ibid.
93. Collins, *Development and Training of the South Vietnamese Army*, 50; Montague, *Advising in Government*, 11.
94. Ibid.
95. Nagl, *Learning to Eat Soup With A Knife*, 153.
96. National Security Council, "Memorandum for Dr. Kissinger: Net Assessment of North Vietnamese and South Vietnamese Forces," (April 19, 1972), https://www.cia.gov/library/readingroom/docs/LOC-HAK-451-2-25-1.pdf, iv–7.
97. Collins, *Development and Training of the South Vietnamese Army*, 12; Krepinevich, *The Army and Vietnam*, 22; Nagl, *Learning to Eat Soup With A Knife*, 119.
98. The major departments of the RVNAF were created: VNAF-25 June 1951, VNN- 6 Mar 1952, VNMC- 13 Oct 1954, Civil Guards and Self-Defense Corps 1956.
99. Collins, *Development and Training of the South Vietnamese Army*, 22.

100. Central Intelligence Agency, *Net Assessment of North Vietnamese and South Vietnamese Military Forces*, 8; Collins (1974); Krepinevich, *The Army and Vietnam*, 22.
101. Collins, *Development and Training of the South Vietnamese Army*, 7.
102. Boot, *Savage Wars of Peace*; Director of Central Intelligence, "South Vietnam: A Net Military Assessment (2 April 1974)," (1974), https://www.cia.gov/library/readingroom/docs/DOC_0001166747.pdf.
103. "South Vietnam: A Net Military Assessment" (1974): 2.
104. Honn, et al., "A Legacy of Vietnam"; Montague, *Advising in Government*.
105. Collins, *Development and Training of the South Vietnamese Army*, Appendix D.
106. Ibid., 81.
107. Krepinevich, *The Army and Vietnam*, 251.
108. "The Vietnam War: Military Statistics," https://www.gilderlehrman.org/history-by-era/seventies/resources/vietnam-war-military-statistics.
109. Collins, *Development and Training of the South Vietnamese Army*, 89–90; Nagl, *Learning to Eat Soup With A Knife*, 174.
110. Nagl, *Learning to Eat Soup With A Knife*, 174.
111. The 196th Light Infantry Brigade and the 3rd Brigade of the 1st Cavalry Division.
112. Central Intelligence Agency, *Net Assessment of North Vietnamese and South Vietnamese Military Forces*, 8.
113. Ibid., iii.
114. Ibid.
115. Ibid.
116. Ibid.
117. Krepinevich, *The Army and Vietnam*, 21.
118. Collins, *Development and Training of the South Vietnamese Army*, 78.
119. Nagl, *Learning to Eat Soup With A Knife*, 170.
120. Collins, *Development and Training of the South Vietnamese Army*, 61–65.
121. Assumes 73,000 per year for desertions from 1959–1964, 113,000 in 1965 and 1966, and 79,100 from 1967–1974 resulting in an estimate of 1,297,000 total estimated deserters from 1959–1974. Conscription evaders is based on 232,000 from 1959–1965 or every six years with no sign of abating. Therefore, I estimate that roughly 696,000 conscription evaders could have possibly refused service in the RVNAF from 1959–

1974 from Collins, *Development and Training of the South Vietnamese Army*, 56–63.
122. 209,517 who were charged and 360,000 who went uncharged, of 213,000,000 US citizens from Andrew Glass, "Carter Pardons Draft Dodgers Jan. 21, 1977," (January 1, 2008), http://www.politico.com/story/2008/01/carter-pardons-draft-dodgers-jan-21-1977-007974.
123. Among these, 44% said Vietnamese forces were generally competent and 40% said they were highly competent.
124. For discussions on American leadership traits, principles, and warfare concepts, see notes 100–102 in Chapter 3, *Advisor Surveys*.
125. Krepinevich, *The Army and Vietnam*, 24.
126. Matthew C. Brand, *Airpower and the 1972 Easter Offensive* (Fort Leavenworth, KS: US Army Command and General Staff College, 1997); John A. Doglione, Donald T. Hogg, Richard D. Kimball, Julian R. McFadden, John M. Rapp, Ray Y. Walden, Lorenz F. Wustner, Charles W. Bond, Eugene T. Buckner, Norman Edgar, *Airpower and the 1972 Spring Invasion* (New York,: Arno Press, 1978); Moore, "North Vietnamese Army's 1972 Eastertide Offensive."
127. Brand, *Airpower and the 1972 Easter Offensive*.
128. Doglione, et al., *Airpower and the 1972 Spring Invasion*.
129. Moore, "North Vietnamese Army's 1972 Eastertide Offensive'"

Chapter 6

State-centric Counterinsurgency

> Occupation duties are the inevitable result of most offensive operations. We need to recognize that a military unprepared for occupation is likewise unprepared for offensive operations. The decision to conquer comes with the responsibility to govern...
> –David A. Mueller (LtCol, USMC)[1]

This chapter will begin with a latitudinal comparison across all four case studies to isolate the principal drivers associated with increasing SLAW. I will examine the evidence for each possible driver—COIN methods, troop ratios, presence of insurgent sanctuaries, continued US support after withdrawal, and degree of embeddedness employed. I will leverage this evidence in paired analysis of case studies of tabula rasa and existing governance. Using these comparisons, I will demonstrate what support there is for my theory of state-centric COIN and conclusions associated with this. Finally, I will examine the potential utility, operability and implications of a theory of state-centric COIN.

Comparing Possible Explanations of SLAW

In this book I identified five compelling potential drivers to explain variations in SLAW in the course of US counterinsurgency interventions. Based on the assessment of the five variables across the four cases, the strongest correlation was between the degree of embeddedness used to develop host-nation security forces and SLAW (see table 11). Where the highest degree of embeddedness was only used solely to develop the host-nation security forces, SLAW was still significant. This indicates that the minimum requirement for increasing SLAW is an effective host-nation security force, even when the host-nation government is authoritarian or dysfunctional. And given the poor showing of strategies of a low degree of embeddedness (i.e., advise and assist), I find that the optimal strategy for developing host-nation security forces in a revolutionary timeframe is strategies of high embeddedness (i.e., encadrement).

The next strongest correlation was between SLAW and the degree of embeddedness used to develop governance and security. When high embedded strategies were employed to develop both governance and security, this resulted in not only a stable security environment for the longest duration, but also improved governance and a stronger democracy.

A conditional correlation was identified between continued US combat support after withdrawal and SLAW. This correlation only applied in strategies of low embeddedness. The use of low embedded strategies resulted in an entirely dependent host-nation government, or strategic rentier state. These strategic rentier states were able to survive as long as strategic rents continued to be paid ad infinitum.

State-centric Counterinsurgency 229

Table 11. Analysis of Competing Explanations for Increases in SLAW.

	Population Centric COIN[1]	Enemy Centric COIN[1]	Troop ratios adequate	Insurgent sanctuaries foreclosed	Combat support for survival?	Degree of Embeddedness (DoE) in governance	DoE in security force	SLAW over 10 years?
Philippines	P, W, E	P, W, E	N-(1:100, 1:494)	No	No	High	High	Yes
Nicaragua	P	P, W	N-(1:100, 1:264)	No	No	Low	High	Yes
Iraq	P, W, E	P, W	Y-(1:70, 1:493)	No	Yes	Low	Low	No
Vietnam	P, W, E	P, W, E	Y-(1:15, 1:32)	No	Yes	Low	Low	No
Correlation to SLAW?	No	No	No	No	Yes*[2]	Yes*[3]	Yes	

The strongest correlation among these variables appears to be high DoE in security force development. At a minimum, increases in SLAW requires high DoE strategies in security force development.

[1] P-present; W-weighted; E-effective.
[2] Conditional correlation—when the US used low embedded strategies, this created strategic rentier states that were dependent on perpetual US support for continued SLAW.
[3] Conditional correlation—the Nicaragua case proves that high DoE strategies in developing governance are not an essential component of SLAW.

The other variables—COIN methodology, troop ratios, and presence of insurgent sanctuaries—did not have positive correlations between their presence and an increase in SLAW. They are relevant tactical and operational subjects, but they are also subordinate concerns for third parties to COIN who ultimately intend to leave.

Population-centric and enemy-centric COIN methodologies failed to explain variations in SLAW. Population-centric and enemy-centric COIN methodologies were evaluated to determine if they were present, weighted, and/or effective.[2] In one case, Nicaragua, while both methods (enemy-centric and population-centric COIN) were present, neither was demonstrably effective. Yet this appears to have had no impact on SLAW as the host-nation government nonetheless persisted over four decades after US withdrawal. The Nicaragua case study was particularly intriguing because it seems to indicate that the United States as a third-party to COIN could fail in its COIN efforts and still increase SLAW as long as it creates an effective host-nation security force.

In contrast, in Iraq, enemy-centric and population-centric methods were present and weighted, yet only the population-centric method could be considered effective as a result of the surge. Nevertheless, this combination of success and failure did not appear to have any measurable impact on SLAW. Furthermore, in Vietnam, both enemy-centric and population-centric methods were present, weighted, and effective. The VC were removed as a relevant military threat by 1968,[3] and, as the Hamlet Evaluation System reported, and Boot supports, as much as 93% of South Vietnamese hamlets had been pacified by 1972.[4] Still, this also appears to have had no discernable effect on South Vietnam's SLAW. Therefore, there appears to be no direct correlation between COIN methods and increases to SLAW.

Counterintuitively, I found that the cases with the highest troop-to-population ratios had the worst outcomes with respect to SLAW. The outcomes appear to be the inverse of what would be predicted by doctrine and theory. The 2006 Field Manual 3-24, *Counterinsurgency*, and

State-centric Counterinsurgency   231

Quinlivan advocate for a 1:50 minimum counterinsurgent-to-population ratio.[5] McGrath is more liberal estimating that the optimal number might be possibly as high as 1:91.[6] The only cases to get within this range were Iraq and Vietnam.

Iraq, at its peak US deployment, was equidistant between the two estimates (1:70).[7] Surprisingly, in Vietnam's case, there were already theoretically more than enough US and South Vietnamese forces on the ground in Vietnam before the first US combat formations intervened in 1965 (table 11). Furthermore, Vietnam had the best troop-to-population ratio of all the four case studies getting as low as 1:15, or one counterinsurgent (US or Vietnamese) for every fifteen South Vietnamese citizens.[8] If it were possible for troop ratios to directly improve SLAW, Vietnam and Iraq should have been the cases to observe the largest increases. Instead, what I observed was that where the Philippine and Nicaraguan cases had the lowest troop-to-population ratios, they ironically fared the best with respect to SLAW.

Similarly, this book found no correlation between the enduring presence of insurgent sanctuaries and SLAW. In fact, without exception, insurgent sanctuaries were present and active before, during, and after US interventions in every case. Counterintuitively, the presence of insurgent sanctuaries in the Philippines, Nicaragua, and Iraq benefitted the United States and host-nation governments more than the insurgents. In all three cases, the insurgents voluntarily cut themselves off from the bulk of the population they meant to influence and control. Even as the sanctuaries, especially in the case of Nicaragua, allowed the insurgents to survive, they also kept them from succeeding.[9] US and host-nation forces had some successes in conducting raids or spoiling attacks into these sanctuaries in all three cases, but they never foreclosed on any of them entirely.

The singular point in mitigation of this observation was in Vietnam. Sanctuaries were important in Vietnam, but not essential to the survival of the VC. From 1965 to 1968, Krepinevich and the CIA both estimated that more than 50% of the VC's support came not from neighboring states or North Vietnam, but directly from the South Vietnamese populace.[10] Further, in 1968–1972, VC and NVA displacement to sanctuaries in neighboring states worked to help, not hurt, pacification efforts. Therefore, in regard to counterinsurgency efforts exclusively, VC displacement to neighboring states helped, not hurt, pacification efforts. The same cannot be said when the war shifted closer to conventional warfare after 1968.

After 1968, the NVA used neighboring states as staging areas for conventional invasions. These conventional invasions were existential threats to the South Vietnamese government's existence. These staging bases are not specifically relative to a discussion of insurgent sanctuaries but rather a conventional invasion from a neighboring state. US commanders addressed these invasion staging bases by conducting spoiling attacks in Laos and Cambodia. The South Vietnamese government could have conducted continued spoiling attacks—preemptive operations designed to hit the enemy while they are still in their staging bases and in a weak position to attack or defend at that time as the US had in its invasions of Laos and Cambodia—to disrupt North Vietnamese invasion preparations. Or it could have invaded the North. Instead, South Vietnam allowed the staging bases to endure unmolested after the US withdrew and suffered as a result. Given these realities, the presence of insurgent sanctuaries does not provide a direct correlation between increases or decreases in SLAW.

The impacts of continued combat support after withdrawal were conditionally corelative. Where the US used institution-influencing strategies, it created strategic rentier states in Vietnam and Iraq; that is, Vietnam and Iraq were able to defend themselves successfully only when they perpetually received strategic rents from the US in the form of combat support (offensive air support, assault support, and intelligence) and combat-service support (logistics). Economic aid was not an issue in either case, as South Vietnam continued to receive economic aid in the amount of $6.03 billion throughout the fiscal year in which they were overrun (in 2019 dollars).[11] Similarly, Iraq received over $860 million in all forms of financial aid in the year before and year of ISIS's 2014 invasion.[12]

State-centric Counterinsurgency   233

Nevertheless, the provision of these titanic amounts of aid did nothing to prevent both countries from being easily defeated after US withdrawal.

Degrees of Embeddedness in Paired Historical Case Studies
This book relied on paired contextually constrained historical comparisons to develop and test a theory of state-centric counterinsurgency. I paired cases where the United States was presented with tabula rasa host-nation governance with cases where a host-nation government existed but was ineffectual. In both sets of contextually constrained historical comparisons, a case of a higher embeddedness was compared to a case of lower embeddedness.

Case Studies of Degree of Embeddedness in Tabula Rasa Governance
The Philippines (1898–1913) and Iraq (2003–2010) provide the first pair of contextually constrained historical comparisons. These represent cases where US intervention created or helped create a context of tabula rasa host-nation governance. The Philippines represented the case of high embedded strategies, and Iraq represents strategies of low embeddedness. The results with respect to SLAW are definitive. Iraq was only able to maintain empirical sovereignty over the majority of its juridical territory for a little over three years after the last US combat formations withdrew in late 2010. It is unclear how much more territory Iraq might have lost if the US had not re-intervened in 2014. In contrast, the Philippine government was able to defeat internal insurgents for a minimum of seventy-one years, and possibly as long as 104 years, after the last US combat formations ceased operations in 1913.

The divergent methods the United States used to develop governance in these cases of tabula rasa governance also produced divergent results. In the Philippines, the United States developed governance symbiotically and transitioned it organically, from the bottom-up. This government lasted seventy-one years after independence and eighty-four years after becoming a commonwealth. In Iraq, the US developed and transitioned governance artificially from outside the institutions and from the top-down. The Iraqi government lost a considerable swath of its territory to a ragtag terrorist group in a little over three years. In the Philippines, the United States made no pretense of host-nation sovereignty initially but promised long-term independence from the start. The United States governed initially at the national level as a military government, then as a civilian trusteeship, and finally through a shared sovereignty arrangement at the provincial and local levels. At the local levels, Filipino leaders made policy and US agents had veto authority. The United States transitioned Filipino governance from the bottom-up through local councils, then magistrates, then a working judiciary, then a permanent national judiciary, then a legislature, and finally a commonwealth and independence.

In Iraq, the US rushed to present a pretense of Iraqi sovereignty, even as the Baghdad government had no budget, no constituency, no safe access to their own citizens, no military, and no extractive means to collect taxes. The United States governed nationally through a figurehead client, or *kafil*,[13] who was compelled to rubber stamp US decisions, as US forces operated unilaterally killing rebellious Iraqi citizens on behalf of the Iraqi government. The US created a hollow Iraqi Governing Council almost entirely of expats, then an Interim Iraqi Government consisting of the same people. The US repudiated local governance, and, rushing to present the façade of Iraqi sovereignty, the US held elections with 25 percent of the population boycotting. The US created a head of government that lacked any body. Any pretense of sovereignty of the Iraqi government was forfeit with the reality that US forces, operating unilaterally on Iraqi soil, were responsible to battle and detain rebellious Iraqi citizens in the name of the Iraqi government.

The divergent methods the US used to develop security forces in a tabula rasa context also produced divergent results. In the Philippines, the US developed security forces symbiotically and transitioned them organically, from the bottom-up. These Filipino security forces would later provide security through several major insurgencies and have endured for over a century. In Iraq, the US developed and transitioned security forces artificially from the top-down, and yet these Iraqi forces could not defend themselves in Mosul despite 30:1 odds in their favor, operating in a defensive posture, holding complex urban terrain, and a massive fire power advantage. ISIS beat the Iraqi Army with around 1,500 fighters arriving in pickup trucks and only 21,000–200,000 other fighters fighting on multiple fronts in Iraq and Syria.[14] ISIS routed two infantry divisions, or 30,000 Iraqi soldiers, and seized the city of Mosul with over 650,000 residents. At the time, Iraqi forces altogether numbered nearly 300,000 active duty soldiers and over 500,000 reservists with modern tanks, aircraft, and artillery.[15] It took the United States and the Government of Iraq years to take back the terrain that ISIS seized in days.

In the Philippines, lessons accrued to the Philippine security forces from within their own institutions. In contrast, Iraqi security forces inherited American lessons and adaptations which were based on unilateral American technological and economic capacity and American military culture. When the Philippine forces were tested, they defeated Communist and Islamic insurgencies without US combat support. They relied on shared lessons and adaptations earned over fourteen years led by US officers. When the Iraqi forces were tested, they collapsed until US combat support was re-introduced.

Iraqi and Filipino security forces were both initially riven by sectarian strife. In Iraq, the sectarian strife was based on the centuries-long rivalries between Sunni and Shiite sects. This distrust was exacerbated between the Sunni officers and Shiite politicians after the US withdrawal. The Philippine military was also beset by ethnic and religious hatreds between the Tagalogs, Macabebes, and Muslims that were as visceral as any in Iraq. Yet, these diverse Filipino troops were able to be amalgamated into a coherent fighting force, led initially by American commanders and later by their own commanders. In contrast, the United States was not able to coalesce the Iraqi security forces and government into a unified force. Instead, through its uniquely American microsectarian political initiatives, the US not only maintained the sectarian divisions but enshrined them in the Iraqi Constitution.[16]

Case Studies of Degree of Embeddedness in Existing Governance

Nicaragua (1927–1933) and Vietnam (1965–1973) provide the second pair of contextually constrained historical comparisons. These represent cases where the US intervened as a result of state frailty but not complete failure. Nicaragua represents the case of high embeddedness to develop the host-nation security forces but low embeddedness to develop the host-nation government. Vietnam represents strategies of low embeddedness to develop both. The results with respect to SLAW are definitive. South Vietnam was completely overrun and ceased to be a country a little over three years after US combat formations withdrew. In contrast, Nicaragua was able to maintain internal stability for forty-six years.

The US developed both Nicaraguan and Vietnamese governance from the outside-in using coercive influence. In both Nicaragua and Vietnam, the US feigned the recognition of host-nation sovereignty. This led to examples of sovereignty fallacy in both cases. First, in both cases, the use of US combat formations to liquidate rebellious citizens at the behest of the Nicaraguan and South Vietnamese governments demonstrated an elemental abdication of host-nation sovereignty. Second, in Nicaragua, the United States chose the US Marines as the military branch to train the Guardia Nacional, despite Nicaraguan preference for US Army trainers. Third, due to the exigencies of the crisis in 1926–1927, US Marines were compelled to operate in combat and lead Guardia Nacional personnel before receiving official legal proviso from the Nicaraguan government to do so. Fourth, where American and Nicaraguan forces operated in the same battlespace, US forces would command both. And fifth, with the exception of paying the Guardia Nacional's budget, due to concerns with graft, the Nicaraguan government would have little say in how its military's budget was spent.

State-centric Counterinsurgency 237

Similarly, in Vietnam, the US imposed an American military model on South Vietnam's armed forces, even though the Saigon government preferred a French-style, territorially based command structure. Second, when President Diem wanted to increase the size of his own armed forces, President Kennedy's permission was required. Kennedy refused to give Diem permission to increase the size of Diem's own military. Third, when President Nixon was restricted by the US Congress from using American forces in Laos in 1971, he was nonetheless able to direct the raid using exclusively South Vietnamese forces.

The US made no concerted effort to develop host-nation government in Nicaragua. In Vietnam, however, the United States relied on CORDS and the District Advisory Program to develop Vietnamese governance through influencing relationships. In Nicaragua, the US sought to create an apolitical military as a check to a partisan and kleptocratic government. US diplomats assumed that an apolitical military was a cause of good governance and not an effect of it. The successes of the District Advisory Program in South Vietnam provided temporary connections between the provincial level and district-level governments. US military advisors provided US military radio connections, transportation, and logistical support that had a valuable impact on governance—but only while the advisors were present. Thus, permanent improvements to dysfunctional South Vietnamese governance at the district level did not occur. At the provincial and national level, CORDS did much to address pacification in 1967–1973. However, CORDS did little to address the systemic corruption that enfeebled the South Vietnamese government. Moreover, the South Vietnamese abandoned the CORDS program immediately following US withdrawal in 1973—a clear rejection of US methods and culture.

The principle difference in the SLAWs of these two cases results from the divergent methods used to develop the host-nation security forces of the two countries. The US employed an institution inhabiting strategy —encadrement—to develop the Guardia Nacional. In contrast, the US employed an influencing strategy, advise and assist, in Vietnam. The

Nicaraguan military adapted like a hybrid American-Nicaraguan force. However, this force was constrained by the economic, technological, and industrial tolerances of Nicaragua.

In contrast, the South Vietnamese military was a near facsimile of the modern (for that day) US military, except for four key areas: 1) the RVNAF rejected American command and control philosophy and culture; 2) the RVNAF suffered from corruption that denuded the capacity of the logistical system; 3) the RVNAF lacked the substantial and effective offensive air support it had come to rely on from the United States; and 4) the RVNAF lacked a national economic and industrial base to support American style operations.

Support for State Centric Counterinsurgency Theory
What I found through these four case studies supports my theory of state-centric counterinsurgency. I found that—when the United States embedded its officers deeply and inhabited host-nation institutions, effecting change from inside host-nation institutions (from inside-out), then state longevity improved in the course of third-party COIN interventions. Conversely, when the United States attempted to effect systemic change in host-nation institutions through influence alone (from outside-in), then state longevity suffered in the course of third-party COIN interventions. As a result, I argue that the most appropriate focal point for third-party counterinsurgents who intend to eventually leave is the state itself.

Therefore, even though focusing on the defeat of the insurgents or the control or security of the population are important tactical concerns, there are not viable strategies for third-party counterinsurgents. In looking at these four case studies I find that no amount of success in the US winning the populace or defeating the insurgents can compensate for failure to develop the host-nation's ability to govern and secure itself after US forces have withdrawn. Therefore, the host-nation government itself, not the population and not the insurgents, is the most appropriate

State-centric Counterinsurgency

focus for US third-party interventions in foreign COIN efforts. This is not to say that population-centric and enemy-centric methods, really hybrid COIN methodologies, are not important. Instead, I find that for a third party who intends to leave, these concerns are subordinate to the primary concern of developing an effective host-nation government and security force.

Firstly, it is important to restate the antecedent conditions under which my theory would be operable. These two antecedent conditions are:

1. The perception or reality of state failure;
2. The perception that only US combat formations are able to arrest this failure.

This is important because, as stated earlier, the investment of US combat formations, operating lethally and unilaterally to help kill the rebellious citizens of a supported state is a radical departure from normal state-to-state intercourse. In fact, as previously mentioned, it has only happened ten times out of the over 300 instances of US use of force abroad.

This book weakens many preconceived notions about COIN. Firstly, I find that it is possible to fail in population-centric and enemy-centric methods and nevertheless succeed in producing a state that is able to survive for decades after US withdrawal—as seen in the Nicaragua case study. Conversely, it is also possible to succeed in population-centric and enemy-centric COIN methods while US forces are present and nevertheless fail to produce a state capable of surviving for more than four years after US withdrawal—as seen in the Vietnam case study. Further, I find that it is also possible to exceed even the largest troop-to-population ratio estimates and still fail to produce a host-nation capable of surviving more than four years after the US forces depart—as demonstrated in the Vietnam and Iraq case studies. Similarly, it is also possible to fall well short of even the most bare-bones troop-to-population estimates and yet succeed in producing a host-nation that is able to survive for decades after US forces leave—as observed in the Philippine and Nicaragua case studies.

Summarized Conclusions

As regards a theory of state-centric COIN and how the degree of embeddedness employed to develop the host-nation's governance and security institutions impact third-party COIN interventions I found that:

1. Bringing about evolutionary change by third-parties can only be accomplished in revolutionary timeframes from within the host-nation institutions themselves.
2. The minimum requirement for increasing SLAW is a capable and effective host-nation security force.
3. When highly embedded strategies were used exclusively to develop host-nation's security forces, security and SLAW increased but not necessarily democracy or good governance.
4. Where US officers deeply embedded in both the host-nation's government and security forces, SLAW, democracy, and governance improved.
5. When the US uses low embedded strategies to develop host-nation institutions, this produces strategic rentier states that suffer from moral hazard such that there is greater incentive for it to perpetuate its own instability to guarantee the perpetual provision of strategic rents; The more assistance that is given, the less effective the host-nation becomes.[17]
6. Strategic rentier states later become crumbling states once the strategic rents they have become accustomed to are withheld.
7. Highly embedded strategies resulted in symbiotic adaptation as US officers effected change from inside the host-nation institutions. These adaptations were neither entirely American nor wholly host-nation. The adaptations were unique to the host-nation's discrete context and became the property of the host-nation institutions within which they were created.
8. Low embedded strategies resulted in artificial adaptation as US officers attempted to effect change from outside the host-nation institutions through influence alone. The unilateral adaptations the US officers transferred were entirely American and were in no

State-centric Counterinsurgency

way connected to the host-nation's ability to sustain them without perpetual US support.

9. Low embedded strategies produced retarding effects on host-nation security forces when US forces shielded anemic host-nation forces from both destruction and learning/adaptation. Low embedded strategies also produced training effects on insurgent forces as US forces and insurgents were compelled to continuously adapt to remain relevant on the battlefield. The US left behind a better trained insurgent force while at the same time retarding the growth of the host-nation forces resulting in a capacity gap between insurgent and host-nation forces.

10. When the US transitioned control back to host-nation institutions in a bottom-up fashion (or grassroots to national level), this increased sustainability and increased the chances of merit-based promotions for host-nation officers.

11. When the US transitioned control back to host-nation institutions in a top-down fashion (national level down), this increased the power of political elites/strongmen to choose officers on the basis of political reliability rather than merit.

12. When highly embedded strategies were employed, US officers were able to act as neutral leaders and unify host-nation security forces that were previously riven by sectarian, religious, ethnic, or ideological strife in ways that non-neutral host-nation officers could not.

Policy Implications

> There was recognition and acceptance of a reality that seems to have been forgotten since, as documented in the Army's history of the occupation of Germany: "Military government, the administration by military officers of civil government in occupied enemy territory, is a virtually inevitable concomitant of modern warfare."
> –David Johnson, *Doing What You Know: The United States and 250 Years of Irregular War*[18]

Whereas institution-inhabiting strategies were effectively employed from 1899 until just after World War II, are they viable options today? The answer is yes. To understand why they are viable today requires addressing political will, probability of success, and political tenability. Both Mao and Clausewitz agreed that war was essentially an extension of politics, or politics by other means. They also agreed that political demands constrained military means.[19] Therefore, determining if the political will exists to apply highly embedded strategies is crucial before nominating it as a potential policy.

Key aspects in determining sufficient political will are duration, cost, and the probability of success of an intervention. With respect to duration, the use of strategies of high embeddedness have resulted in conflicts of shorter duration and exponentially longer SLAWs. The average duration of combat operations by US forces in all of its conventional counterinsurgency interventions has been approximately a decade (8.85 years before World War II and 10.6 years after; see table 1). However, the duration of US support to counterinsurgencies since World War II has dramatically increased. The United States led the training and support for South Vietnamese COIN operations for nineteen years (1954–1973). The US has been leading the counterinsurgency support for the Afghan state for the last eighteen years (2001–present) and the Iraqi state for sixteen years (2003–present). The average of these COIN support missions after World War II, if they were to end today, would be around sixteen years. However, US interventions in Iraq and Afghanistan have no end in sight. In contrast, the Philippines took fifteen years to conclude decisively, no longer requiring US combat forces or combat support, and Nicaragua took only five. The US remained in the Philippines to provide American access to Far East markets, not to keep the Philippine state alive.

All these wars, before and after World War II, were costly in terms of blood and treasure. All of these wars were incredibly complex. Therefore, the principle differences that set them apart are their durations and probabilities of success. The results are conclusive: the use of highly

State-centric Counterinsurgency

embedded strategies result in interventions of shorter duration and produce states that last far longer after cessation of US combat operations. What about domestic political tenability?

Are Highly Embedded Strategies Politically Tenable Today?

> While the army's embrace of military governance may appear strange in 2016, the US Army of 1940 could refer to a long list of precedents in which US occupation required military governments: the Reconstruction following the end of the Civil War in 1865, the Philippines 1898–1946, Cuba 1898–1902, Puerto Rico 1898, Veracruz, Mexico 1914, the Rhineland 1918–1923, and numerous Marine Corps interventions in the Caribbean. Together, these occupations represent more than a hundred twenty years of consistent, periodic, need for military governments.
> –Lt Col David A. Mueller (USMC)[20]

While institution-inhabiting strategies were politically tenable from 1899 until the aftermath of World War II, are they tenable today? The answer is yes. The key arguments against the tenability of highly embedded strategies today are: 1) they appear neo-imperial/neo-colonial;[21] 2) host-nation populations will reject control by a foreign power;[22] and 3) practical concerns over language issues and force protection make them infeasible.[23] The claims over force protection and language are most easily addressed as these concerns manifest themselves whether the US employs high or low embedded strategies.

Regarding the appearances of neo-imperialism/neo-colonialism, there are several reasons to reject these critiques with respect to modern neo-trusteeships and shared sovereignty arrangements. Claims of neo-imperialism are invalidated by the inherent differences between classical imperialism and neo-trusteeships, the dilution of claims of neo-imperialism, and the charge of neo-imperialism associated even with the current methods. James Fearon and David Laitin point out the significant differences between classical imperialism and modern neo-trusteeships.

Although they concede that both share a prolific degree of control by foreign powers over internal host-nation affairs, they argue that classical imperialism and modern neo-trusteeships differ fundamentally. This difference exists because a host-nation under neo-trusteeships is "governed by a complex hodgepodge of foreign powers, international and nongovernmental organizations (NGOs), and domestic institutions, rather than by a single imperial or trust power asserting monopoly rights within its domain...the parties to these complex interventions typically seek an international legal mandate for their rule...whereas classical imperialists conceived of their empires as indefinite in time, the agents of neo-trusteeship want to exit as quickly as possible..."[24]

Another argument for the tenability of highly embedded strategies is that the US's reliance on less invasive strategies have nonetheless failed to shield the US from claims of neo-imperialism. The move away from highly embedded strategies toward less invasive strategies was a direct reaction to charges of colonialism and imperialism during the Philippine and Banana Wars as well as during the Cold War, and Global War on Terror. However, this shift has not averted continued charges of neo-imperialism and neo-colonialism even against the current use of less invasive, low embedded COIN strategies such as advise and assist.[25]

There appears to be little to be gained by using less invasive strategies with respect to defending against claims of neo-imperialism. Claims of neo-imperialism and neo-colonialism has continued to perpetuate since 1950 despite the exclusive use of low embedded strategies. Moreover, the over-profusion of charges of neo-imperialism have diluted the gravity of such claims. What exactly constitutes imperialistic behavior has become diffused by claims of economic imperialism,[26] democratic imperialism,[27] humanitarian imperialism,[28] developmental imperialism,[29] linguistic imperialism,[30] and cultural imperialism.[31] Claims of neo-imperialism will exist as long as the US maintains a power imbalance with other states and continues to intervene in internal conflicts of other states. There appears to be little benefit to choosing less invasive strategies as a means of deflecting charges of neo-imperialism.

Therefore, if any form of US intervention is likely to be labeled as neo-imperialistic, it seems to make more sense to choose a strategy with a higher probability of success and historic lower costs. It also makes sense to choose a strategy that is at least as politically viable and that possesses a higher probability of success. It appears to make less sense to choose a strategy with less probability of success and higher historic costs that will nevertheless be impugned as neo-imperial either way.

The argument that host-nation populations will automatically reject foreign forces leading their security institutions or administering their governance institutions is not new—neither is it well supported by history. General Edmund Allenby in Syria cautioned General Harry Chevelle not to allow British troops to enter Damascus in 1918. Allenby's warning was out of fear of an uprising caused by Muslim peoples opposed to Western occupation. Similar concerns led the Bush Administration and General Tommy Franks to keep US forces outside major Iraqi cities in 2003.[32] However, the Ottoman government abandoned Damascus in 1918, and Iraqi security forces abandoned their posts throughout Iraq in 2003. Chevelle was compelled to enter and secure Damascus in much the same way the United States was compelled to provide security and governance in the absence of Iraqi security and governance. The principle complaint from this early period in the Iraq War was not that the United States had too many Western occupiers, but too few.[33]

It is specifically due to the recorded failures of low embedded, institution-influencing strategies that makes high embedded, institution-inhabiting strategies viable once again. A valid critique of the contemporary tenability of highly embedded strategies is that if they were expressly rejected in Vietnam and Iraq, what makes them more viable after these conflicts? I argue that it is expressly due to the contemporary military/political contexts created by the failures in Vietnam, Iraq, and potentially Afghanistan that make strategies of high embeddedness more tenable today than they were only fourteen years ago. If it were possible to win a foreign counterinsurgency and create a state that would endure after US withdrawal through massive economic investment and large numbers of US troops alone, then US efforts in Vietnam and Iraq should have produced states lasting far longer. However, large-scale COIN interventions in Vietnam and Iraq have only produced states capable of surviving less than four years after US withdrawal. In contrast, large-scale COIN interventions have increased SLAW dramatically when inhabiting strategies were used to develop host-nation institutions.

Lastly, are high DoE strategies politically tenable today? They can be. This is demonstrated most clearly by the fact that they are already being used today. Cindy Daase, Fearon and Laitin, Mueller, and General Zinni provide examples where high embedded strategies are being employed today and guidelines on how they could be employed in the future.[34]

Daase, Caplan, Johnson, and Mueller have not only argued for the value of strategies of high embeddedness, but they have identified where they are already being employed or where preparations are being made for their employment.[35] Richard Caplan notes that, "Impracticable though the idea of international trusteeship may have seemed to many at that time, it soon became a reality in all but name with the establishment of international territorial administrations in Bosnia and Herzegovina (1995), Eastern Slavonia (1995), Kosovo (1999) and East Timor (1999)."[36] Cindy Daase does extensive research on the efficacy of shared sovereignty arrangements employed in developing Sierra Leone and Liberia from 2005–2006.[37] Finally, Johnson notes presciently that, "the Army recognizes the requirement to consolidate gains post-conflict, noting in *The US Army Operating Concept: Win in a Complex World*, 'The Army also prepares for security operations abroad including initial establishment of military government pending transfer of this responsibility to other authorities.' Furthermore, the establishment of the Institute for Military Support to Governance (IMSG) at the US Army John F. Kennedy Special Warfare Center and School in 2013 was an important step in turning concepts into

State-centric Counterinsurgency 247

doctrine and organizational capability."[38] If highly embedded strategies are politically viable, politically tenable, already being used, and come with higher probabilities of success, under what guidelines would they need to operate in order to maintain these criteria?

Fearon and Laitin, Daase, Mueller, and Zinni nominate several guidelines under which high embedded strategies should be employed.[39] Fearon, Laitin, and Daase agree that one of the key criteria for implementation is a clear indication that the appropriation of the host-nation sovereignty is temporary in nature.[40] They and Zinni also argue for the oversight of the foreign intervention force by international organizations, as well as multilateral participation, in order to avoid imperialist and neo-colonial criteques.[41] In contrast, Mueller argues that, based on the lessons from the US occupation of the Rhineland in 1918, the US Army determined that highly embedded strategies require strict adherence to unity of command.[42] This does not necessarily obviate the former arguments of Fearon, Laitin, Daase, and Zinni. It is possible that there could be a single foreign commander, civilian or military, who is in charge, but who is also reponsible to an international governing body. Additionally, Daase, reflecting on her comparisons between different instances of shared soveereignty arrangements in Sierra Leone and Liberia, advanced a couple of additional key requirements. She argued that for success, highly embedded strategies need to be "self-enforcing and self-executing" and "have built-in sanctions and for defection and incentives for cooperation".[43]

Transitioning to institution-inhabiting strategies today would require minimal change in the structure and doctrine of the current US DoD. The most elemental change would need to take place at the doctrinal level. The US has documents of historical precedence to rely on, such as *The Small Wars Manual* (1940) and Field Manual 27-5, *Military Government and Civil Affairs* (1940). These historical publications would need to be updated and made relevant to some of the guidelines mentioned by Fearon and Laitin, Daase, and Zinni.

Further, the US military would need to integrate training in military government and encadrement into its professional military education (PME) system. This could occur in several areas. Rebecca Patterson provides a valuable work on how the School of Military Governance trained officers for post-conflict World War II occupation and rebuilding.[44] Some of this structure already exists in the US military. Both the US Army and the US Marine Corps now have standing theater-security cooperation institutions. These could be adapted to prepare officers and NCOs for the challenging responsibilities of encadrement and military government. Additionally, the military PME system could easily add curricula to its formal courses to educate officers and senior NCOs on strategies of high embeddedness.

Lastly, the US would also require a substantial deployable civilian force capable and able to assume the role of trusteeship after a state transitioned from US military government. This has been tried before with the US Department of State's Civil Response Corps (CRC). However, the State Department does not have the capacity to maintain a standing deployable force. It could, however, train a significant number of potential foreign service officers in US military PME schools. Then, these foreign service officers would return to their normal duties until a crisis arose. Their skills, once learned, could then be mobilized in a crisis.

Strategies of high embeddedness are not panaceas in and of themselves. The French employed encadrement in Indochina and Algeria and still failed. In contrast, the US used these strategies and succeeded in increasing SLAW after they withdrew. So, what explains the divergent outcomes as a result of the use of high embedded strategies by American and French forces? The answer lies in the contrasting purposes of the French and Americans. The purpose of French efforts was to create dependency and a permanent French colony. The purpose of US institution-inhabiting strategies has been to create an independent host-nation and to allow US troops to leave. Indeed, this key difference is the same difference between classical imperialism and colonialism and contemporary high embedded strategies like neo-trusteeships. The goal of modern neo-trusteeship relations is to allow the intervening forces to leave quicker, not to stay longer. The French in Indochina were not trying to develop a host-nation security force capable of standing on its own. The French desired a capable, but dependent, local ally. In contrast, American inhabiting strategies before 1950 sought to create a viable independent host-nation that would allow the US to withdraw its troops.

Potential Value of Highly Embedded Strategies— Best Worst Options?

> Their limitations notwithstanding, what may most recommend arrangements of this kind in the future is simply that they can represent the least worst option, if not sometimes the best hope, for easing a territory's transition from war, injustice and dependence to peace, basic human rights and relative independence in the context of the changing international order of the post-cold war era.
> –Richard Caplan, "From Collasping States to Neo-Trusteeship"[45]

Institution-inhabiting strategies may assist intervention planners and policymakers in articulating a more concrete and achievable purpose for intervention. This research suggests that the historical purpose of American high embedded strategies has been to produce viable states during COIN interventions that are able to govern and secure themselves after the withdrawal of US forces.

Winston Churchill once appraised democracy as the worst option for governing a state except for all the others—and so it is with strategies of high embeddedness. Given the poor track record of low embedded strategies, there appear to be few alternatives. One obvious alternative is to stop intervening. As Jeffery Herbst offers, the international system could let weak states die, split, or shrink to the point where their juridical borders match their empirical capacity to govern and secure themselves.[46] The second option is to simply contain the fallout of state failure. This strategy would seek to limit the impacts of state failure on the rest of the international system, but still allow the state to collapse and then rebuild itself organically.

However, if the situation appears so dire that only the intervention of US forces is seen as sufficient to arrest it, then the US has high and low embedded strategies to choose from regardless of which COIN method it uses, its troop ratios, or its efforts to interdict insurgent sanctuaries. Though, given the poor historical record of low embedded strategies once the situation has deteriorated to the point where US formations were required to intervene, these should be considered politically untenable.

Therefore, the relevant argument is not between isolationism and imperialism, but between the most efficacious strategy to remediate state failure in COIN interventions when only foreign combat formations can arrest it. My research does not advocate for the existence of a false dichotomy between isolationism and imperialism in choosing high and low embedded strategies. But other options seem limited. There appear to be only three historical American options once the situation has devolved to the point where only US combat formations are sufficient to repair it:

1. A large-scale COIN intervention and low embedded strategies resulting in strategic rentier states with short SLAWs;
2. A large-scale COIN intervention and high embedded strategies resulting in stumbling or crumbling states, albeit with long SLAWs;
3. Do not intervene, and let weak states fail.

Where the state is not in imminent threat of failure or the situation is not dire enough to require the intervention of US combat formations, then myriad options are available and may be more valuable.

National security guidance contained in the 2012 National Security Strategy, *Sustaining the US Global Leadership: Priorities for 21st Century Defense,* signaled an official departure from large-scale US COIN operations. Specifically, presidential strategic guidance directed that the US military "will no longer be sized to conduct large-scale, prolonged stability operations."[47] Furthermore, there does not appear to be an increasing desire to get involved in large-scale foreign COIN operations. However, General James Mattis, the former Secretary of Defense, has previously noted that "the enemy also gets a vote." Thus, the decision to intervene may be forced upon the US and may not be of its own volition. This book finds several alternative possible outcomes. The first result might be a stumbling state with high stability, but imperfect democracy, like the Philippines. Another possible outcome is the tumbling state, which is a stable but faux democracy at best, such as Nicaragua. The final alternative is to produce strategic rentier states that are able to endure only as long as the US perpetually provides economic and military combat support, such as cases like Iraq and Afghanistan.

As the United States seeks to conclude the wars in Iraq and Afghanistan, it seems clear that, as things stand now, two strategic rentier states have been created. As long as the United States is willing to provide the strategic rents to which the Iraqi and Afghan governments have gotten used, there is reason to assume that they will continue to exist indefinitely. However, based on my research in this book, should the US withhold the training, advisors, and combat support that these states have gotten used to, it is likely that they will collapse within three or four years of support being terminated.

Further, as the United States seeks to preempt instability through US military security cooperation deployments all over Africa and elsewhere, the United States may be compelled, against its druthers, to intervene with combat formations. The addition of more advisors to such a situation will not repair a failed or failing state through influence alone. If the situation has gotten so bad as to require US combat formations to arrest it, dysfunctional or nonexistent host-nation institutions cannot be reformed, rehabilitated, or created from scratch, strictly through influence. Caplan calls neo-trusteeships the best-worst option available. The United States should prepare itself for this contingency or relegate itself to pursuing strategies of low embeddedness. Unfortunately, low embedded strategies are expensive in lives and treasure and provide low historical probabilities of success once a host-nation's situation has deteriorated to the point where only US combat formations can arrest it.

Notes

1. Mueller, "Civil Order and Governance," 49.
2. Please find criteria for presence, weightedness, and effectiveness of COIN methods in Chapter 1 under "Evaluating Possible Explanations for Increasing SLAW-COIN Methods."
3. Boot, *Savage Wars of Peace*, 308; Nagl, *Learning to Eat Soup With A Knife*, 167; Krepinevich, *The Army and Vietnam*, 239.
4. Andrade and Willbanks, "CORDS/Phoenix," 17; Boot, *Savage Wars of Peace*, 315; Hamlet Evaluation Survey (HES) Annual Statistical Analysis (1968–1971); White, "Civil Affairs in Vietnam"; Honn, et al., "A Legacy of Vietnam."
5. Quinlivan, "Force Requirements in Stability Operations"; FM 3-24 (2006).
6. McGrath, *Boots on the Ground*, 109.
7. *Measuring Stability and Security in Iraq* (2006); McGrath, *Boots on the Ground*, 193; Petraeus, *Report to Congress*; Belasco (2009).
8. Central Intelligence Agency (1967); Director of Central Intelligence (1966): 6; Director of Intelligence (1968); Director of Central Intelligence (1969); National Security Council (1972): IV-7; Central Intelligence Agency, *Net Assessment of North Vietnamese and South Vietnamese Military Forces*, 4; Director of Central Intelligence (1974): iii.
9. Millett, *Guardians of the Dynasty*.
10. Central Intelligence Agency (1966); Krepinevich, *The Army and Vietnam*.
11. Ellerman, "Memorandum: The South Vietnamese Economy," 5.
12. USAID (2015).
13. This Iraqi figurehead regime approximates what Luciani and Beblawi term a *kafil* in Arabic societies and in the study of *rentier states*. According to Luciani and Beblawi, a *kafil* is a local sponsor who is employed by a foreign entity to legitimate its independent operations in another state ("Rentier State in the Arab World," 92).
14. "Iraq Army Capitulates to ISIS Militants"; Gartenstein-Ross (2015).
15. "Current Military Capabilities and Available Firepower for 2016Detailed."
16. Jawad, *The Iraqi Constitution*.
17. Beblawi, "Rentier State in the Arab World," 385–386; Krepinevich, *The Army and Vietnam*, 3; Mahdavy, "Pattern and Problems of Economic Development in Rentier States," 437.

18. David Johnson, *Doing What You Know: The United States and 250 Years of Irregular War* (Washington, DC: Center for Strategic and Budgetary Assessments, 2017), 81.
19. Clausewitz, Howard, Paret, and Brodie, *On War*; Mao, *On Guerilla Warfare*.
20. Mueller, "Civil Order and Governance,", 48.
21. Lieutenant General Robert B. Neller, Interview with Author (March 6, 2015); Petraeus, "Counterinsurgency"; Zinni interview (2017).
22. Mueller, "Civil Order and Governance."
23. Nagl, *Learning to Eat Soup With A Knife*.
24. Fearon and Laitin, "Neotrusteeship and the Problem of Weak States," 7.
25. Bill Van Auken and David North, "The Iraq War ten years on: A turning point for US imperialism," (March 19, 2013), https://www.wsws.org/en/articles/2013/03/19/pers-m19.html; Laleh Khalili, *Time in the Shadows: Confinement in Counterinsurgencies* (Stanford: Stanford University Press, 2012), see portions on US counterinsurgency methods and 1–7 and 239–249; Montague, *Advising in Government*, 12; Paul Amar, *The Security Archipelago: Human-Security States. Sexuality Politics, and the End of Neo-Liberalism* (Durham: Duke University Press, 2013), 55, 115–116, and 235–252 discussing new *security dialouges*; Richard Becker, " How Vietnam Defeated US Imperialism," (March 8, 2015), http://liberationschool.org/how-vietnam-defeated-u-s-imperialism/; Robert Shilliam, "What the Haitian Revolution Might Tell Us about Development, Security, and the Politics of Race," *Comparative Studies in Society and History* 50, no. 3 (July 2008): 778–808.
26. Edward P. Lazear, "Economic Imperialism," *The Quarterly Journal of Economics* 115, no. 1 (2000): 99–146.
27. Stanley Kurtz, "Democratic Imperialism: A Blueprint," (April 1, 2003), https://www.hoover.org/research/democratic-imperialism-blueprint.
28. Ray Bush, Giuliano Martiniello and Claire Mercer, "Humanitarian Imperialism," *Review of African Political Economy* 38, no. 129 (August 2011): 357–365.
29. Alfred Zack-Williams, "Neo-imperialism and African Development," *Review of African Political Economy* 40, no. 136 (June 2013): 179–184.
30. Anne Burns, "Is English a Form of Linguistic Imperialism?" (April 10, 2013), https://www.britishcouncil.org/voices-magazine/english-form-linguistic-imperialism.

31. Thomas Auxter, "The Debate over Cultural Imperialism," *Proceedings and Addresses of the American Philosophical Association* 59, no. 5 (June 1986): 753–757.
32. Mueller, "Civil Order and Governance," 47; Bensahel, et al., *After Saddam*, 85–89.
33. Bensahel, et al., *After Saddam*; Mueller, "Civil Order and Governance."
34. Daase, "International Governance via Shared Sovereignty"; Fearon and Laitin, "Neotrusteeship and the Problem of Weak States", Mueller, "Civil Order and Governance"; and Zinni interview (2017).
35. Daase, "International Governance via Shared Sovereignty"; Caplan, "From Collapsing States to Neo-Trusteeship", Johnson, *Doing What You Know*; and Mueller, "Civil Order and Governance."
36. Caplan, "From Collapsing States to Neo-Trusteeship", 231.
37. Daase, "International Governance via Shared Sovereignty."
38. Johnson, *Doing What You Know*, 81.
39. Fearon and Laitin, "Neotrusteeship and the Problem of Weak States"; Daase, "International Governance via Shared Sovereignty"; Mueller, "Civil Order and Governance"; and Zinni interview (2017).]
40. Fearon and Laitin, "Neotrusteeship and the Problem of Weak States," 7; Daase, "International Governance via Shared Sovereignty."
41. Zinni interview (2017).
42. Mueller, "Civil Order and Governance."
43. Daase, "International Governance via Shared Sovereignty."
44. Patterson, *Revisiting a School*.
45. Caplan, "From Collasping States to Neo-Trusteeship," 231.
46. Jeffrey Herbst, "State Failure," in *When States Fail: Causes and Consequences*, edited by R. Rotberg (Princeton, NJ: Princeton University Press, 2003).
47. US National Security Strategy, *Sustaining the US Global Leadership: Priorities for 21st Century Defense* (Washington, DC: National Security Council, 2012), 6.

Index

Abaunza, Gustavo, 158
Abrams, Creighton, 75, 173–175
Abu Ghraib Prison, 118, 130
Academia Militare, 156–157
Act #175 of the 2nd Philippine Commission, 59
adaptation, 16–17, 25–26, 36, 56, 60, 68, 86, 88, 95, 99, 103, 120, 122–123, 153, 212, 235, 240–241
advise and assist, 19, 120, 144, 214, 228, 237, 244
advisor, 15, 63–64, 66, 93–95, 101–102, 107–109, 112–113, 115–117, 120, 123–124, 130, 175–177, 182–185, 187–188, 190–192, 194–195, 197, 200, 203–212, 214, 219, 223, 225, 237, 251
Afghan, xv–xvi, 2–3, 127, 242, 251
Afghanistan, xv–xvi, 1–3, 8, 29, 33, 36, 50, 66, 80, 86, 102–103, 242, 245, 251
Aguinaldo, Emilio, xvi, 41–42, 44–45, 49, 52–53, 56–58, 60, 62, 70, 72
air-mobility, 172
Al-Alwani tribe, 88
Al Anbar, Iraq, xvii, 82, 88, 99, 107, 118–119, 122, 126
Al Assad, Hafez, 9
Al Baghdadi, Abu Bakr, 102, 119
Al Maliki, Nuri, 76, 100
Al Qaeda in Iraq (AQI), xvii, 89, 120, 122
Al Sadr, Moqtada, 81, 101, 126

Al Sahwah, 88
Al Zarqawi, Abu Musab, 89
Al-Alwani, Ahmad, 88, 118
Albu Risha tribe, 88
Algeria, 248
Al-Hawza, 81
Al-Jabbouri, Kadhim Sharif Hassan, 75
Al-Khatteeb, Luay, 93, 128
Allawi, Iyad, 98
Allenby, Edmund, 245
al-Yawer, Ghazi, 98
American, xvi, 2–3, 5–6, 11–12, 24, 26, 34–35, 37, 39–41, 43, 45, 47, 51–61, 64–68, 70–71, 76, 78, 81, 83–84, 87, 94–95, 98–101, 112, 117, 123, 126, 134, 139–140, 143, 145–148, 150, 152–153, 156–158, 167, 171, 173, 175–176, 183–185, 190–193, 195, 197–198, 200–201, 206–207, 212, 215–217, 225, 235–238, 240, 242, 248–250, 255, 279
American foreign policy, 143, 184
American Legation in Managua. *See under* Managua
Amos, James, 15
An Loc, South Vietnm, 213
An Najaf, Iraq, 81, 83
An, Pham Xuan, 189
anarchy, 77–78, 136
Anbar Awakening, 88
antecedent conditions, 17, 28, 239

anti-communist. *See under* communist
anti-Diem coup, 170
Ap Bac, South Vietnam, 170
Arellano, Cayetano, 52, 54
Army for the Defense of Nicaraguan Sovereignty. *See under* Nicaragua
Army of the Republic of Vietnam (ARVN). *See under* Vietnam, 168–173, 181, 185, 189, 193–194, 198, 200, 202, 206, 212–213, 217
Arreguin-Toft, Ivan, 10
artificial adaptation, 26, 122–123, 240–241
Asiatic Squadron, 39
assault support, 14, 76, 200, 232
Audiencia Teritorial de Manila, 54
Austin, Lloyd, 15
Austria, 183
authoritarian, 7, 9–10, 12, 23, 33, 46, 228
authoritarian COIN, 7, 9–10, 46
Axis Powers, 144

Baathist, 80, 82
Baghdad, 75, 79, 89, 98, 107, 118–119, 125–126, 234
Baji. *See under* Iraq
Baliuag, The Philipines, 54
Banana Wars, 244
bankrupt, 136
Basra. *See under* Iraq
Bass, John, 1, 40
Batson, Matthew A., 56–57, 72–73
Battalion Landing Team 3/9, 169
battles
    Battle of Abu Ghraib, 83
    Battle of Al Qaim, 83
    Battle of An Najaf, 81, 83

battles (*continued*)
    Battle of Bud Bagsak, 49, 58–59, 68, 73
    Battle of Bud Dajo, 49, 59, 73
    Battle of Fallujah (1st battle), 82, 105
    Battle of Fallujah (2nd battle), 82, 105
    Battle of Ramadi, 83
    Battle of Samarra, 83
    Battle of Tal Afar, 83
    Battle of Tirad Pass, 44
Beadle, Elias, 151
Beblawi, Hazem, 23–24, 36, 96, 129, 253
benevolent assimilation, 43, 52, 71, 74
biometric identification systems, 87
Blackwater, 82
Boot, Max, 2, 14–15, 34–35, 71, 73, 165–166, 178–179, 219–222, 224, 230, 253
Bosnia and Herzegovina, 246
bottom-up development, replacements, and transitions, 65, 95, 233–234, 241
Boyd, John, 25, 36
Brand, Matthew C., 212, 225
Bremmer, L. Paul, 78–79, 98, 105
British, xvi, 11, 36, 39, 55, 59, 96, 245
Brookings Doha Center, 93
Brown, John S., 90, 176
Buddhists, 170
budget or budgetary control, 93, 121, 146–149, 155, 234, 236
Buencamino, Felipe, 42
Bush Administration, 82, 245
Bush, George W., 40, 70, 82, 99, 245, 254
Byman, Daniel, 9, 33

Index   259

C Company, 7th Brigade Combat
    Team (Philippine Army), 63
Cailles, Juan, 58
Calderon, Felipe, 42, 45, 66
Cambodia, 169, 178, 180, 203, 213,
    232
Camp Bucca, 102
Caplan, Richard, 16, 34, 246, 249,
    251, 255
Capture-kill operations, 87
Carter, Calvin, 148–149, 154, 165,
    225
Carter's Guardia Nacional or
    Carter's constabularies, 148–149,
    154, 165
Castro, Fidel, 10
Catholics, 58, 140, 170
Central Intelligence Agency (CIA),
    63, 168, 175, 177, 179, 181–182,
    184, 189, 193, 195, 198, 200–201,
    218, 220–224, 231, 253
Chamorro, Emiliano, 135–136,
    148–149, 154
Chamorro-Rivas Revolt, 148
charge d'affaires, 137
Chechnya, 10
Chevelle, Harry, 245
Chieu Hoi (Open Arms) Viet Cong
    amnesty program or Hoi Chanh,
    188–189
China, 33, 61, 169–170, 182, 201,
    279
Churchill, Winston, 249
Civil Operations and Revolutionary
    Development Support (CORDS),
    171, 173–175, 186–190, 202,
    219–220, 222–223, 237, 253
Civil War, 46, 63, 84, 86, 134, 139,
    159, 243
civil-military operations, 187

classical imperialism, 243–244, 248
Clausewitz, Carl von, 25, 36, 242,
    254
clear, hold, build, 106
Clifford, Clark, 197
Coalition Military Assistance
    Training Team (CMATT), 105
Coalition Provisional Authority
    (CPA), 78–79, 81, 84, 95–98, 100,
    104, 121, 126
Coalition Provisional Authority
    Order #2, 78–79, 104
Coast Rica, 180
coercive influence, 19, 62, 123, 144,
    160, 183–184, 215, 236
Cold War, 143–144, 169, 244, 249
collateral damage, 9, 87
Collins Jr., James Lawton (Brigadier
    General), 33, 72, 170, 195–196,
    199, 219, 222–225
colonialism, 183, 243–244, 248
    western colonialism, 170
combat operations, 4, 6, 32, 51, 68,
    75, 149, 154, 243
    combat support, xvii, 5, 14–15,
    19, 24, 49–50, 60, 64–65,
    84, 93, 143, 153, 180–182,
    197–198, 201, 208, 213, 217,
    228, 232, 235, 242, 251
    combat-service support, xvii,
    14–15, 24, 60, 64, 153,
    180–181, 232
Combined Joint Task Force 7/
    Multinational Force Iraq, 84
command and control, xvii,
    193–194, 238
command relationships, 145–146
Commandant of the Marine Corps,
    147
Commanders Emergency Response
    Program (CERP), 88

commonwealth, 55, 233–234
communist, 32, 46, 64, 73, 144, 168–170, 172, 174, 189, 198, 216, 235
    anti-communist, 143, 183
    communist aid, 181
    communist critiques, 183
    communist insurgency, 61
    Communist revolutionary war, 190
    non-Communist, 167
compadrazgo system, 22, 65, 67–69
Company M, USMC in Nicaragua, 141
conscript(s), conscripted,, 2, 140, 150–151, 202–203, 216–217, 224
conscription and desertion laws, 203, 216–217
constabulary, 35, 59–60, 69, 73, 133, 135, 148, 154, 163, 165
constituency, 96–97, 121, 170, 234
contextually constrained historical comparison, 28, 233, 236
conventional combat operations, 63, 75, 105, 198, 213
conventional forces or combat formations, 4, 27–28, 179–180, 185, 187
conventional warfare, 44, 70, 137–138, 178, 232
cordon and knock, 87
cordon and search, 87
Corinto, Nicaragua, 133, 135
corruption, 1, 17, 22–23, 54, 76–78, 94, 104, 106, 108, 115–117, 123, 125, 152, 162, 168, 174, 186, 188, 190–192, 195, 201, 206, 208, 211–212, 216, 237–238
Costa Rica, 142–143

counterinsurgencies
    counterinsurgency (COIN), 1–18, 20, 22, 25–30, 32–35, 40, 43–44, 46–47, 58, 62, 64–66, 68, 71, 74, 76, 80, 83–84, 86, 88, 92, 106, 122, 126–127, 130, 137, 139–141, 170–172, 174–176, 178–180, 188, 214–215, 219–220, 223, 227–228, 230, 232–233, 238–240, 242, 244, 246, 249–251, 253–254, 279
    counterinsurgency (COIN) interventions, 1, 4–6, 12, 16–18, 20, 22, 25, 28, 30, 40, 68, 228, 238, 240, 242, 246, 249–250
    counterinsurgency (COIN) methods, 12, 46, 65, 80, 137, 139, 170–171, 174–175, 178, 214, 227, 230, 239, 250, 253–254
    counterinsurgency support, 242
    counterinsurgent, 2, 8–13, 25, 27, 47, 88–90, 92, 139, 141–142, 176–177, 231, 238
    large-scale COIN interventions, 4, 246
crossover point, 172–173, 175
crumbling state, 22, 25, 120, 167, 214, 217, 240, 250
Cuba, 29, 43, 141, 144, 243
cultural imperialism, 244, 255
Cultural Revolution, 10

Damascus. *See under* Syria
Danang, South Vietnam. *See under* Vietnam
de-Baathification, 79
DeBayle, Anastasio "Tachito" Somoza, 139

Index

decapitation strikes, 46, 80, 83
degree of embeddedness, 7, 16, 20–22, 30–31, 227–228, 233, 236, 240, 246
demilitarized zone (DMZ), 14, 179
democracy, 5, 25, 35, 65, 69, 120, 228, 240, 249
   democratic, 22, 66–67, 89, 95, 99–101, 140, 162, 244, 254
   democratic imperialism, 244, 254
   faux democracy, 23, 160, 251
   Western liberal democracies, 9–11
Department of Defense (DoD), 8, 13, 35, 77, 85, 101, 105, 127, 150, 247
Department of State, 77–78, 103, 135, 144, 146–148, 158, 160, 181, 183, 248
Department of the Treasury, 181
Destruction and Creation, 25, 36
developmental imperialism, 244
Dewey, George, Commodore, 39, 41
Diamond, Larry, 95, 99–100, 129
Diaz, Adolfo, 136, 145–146
dictator or dictatorial, 41, 135, 137, 140, 162
   dictatorship of the minorities, 98
Diem regime, 170
Diem, Ngo Dinh, 170, 183–185, 188, 194, 215–216, 219, 237
Dien Bien Phu, 175
District Advisory Program, 174, 191, 222, 237
District Advisory Teams, 191
district chief, 187
district level, 190–191, 237
Dominican Republic, 29, 141
Duggan, Laurence, 158
Dulaimi Tribal Confederation, 118

Dunford, Joseph, 15
duration, 4–6, 47, 69, 83, 156, 162, 182, 228, 242–243

East Timor, 246
Easter Offensive (1972), 15, 203, 212–213, 225
Eastern Slavonia, 246
Eberhardt, Charles, 134
economic aid, 14, 93, 143, 167, 181, 232
economic and industrial base, 193, 216, 238
economic assistance, 93, 168, 180
economic imperialism, 244, 254
effecting change inside-out, 238
8th Corps (US Army Expeditionary Forces in the Philippines), 57
El Salvador, 5–6
election boycotts, 63, 99, 101, 121, 234
elections, 17, 19, 42, 55, 61–64, 76, 84, 89, 96, 99–101, 103, 106, 121, 129, 134, 136, 138, 142–143, 146–147, 149, 151, 154, 157, 164–165, 234
embedded Provincial Reconstruction Teams (ePRTs), 103
embeddedness, 7, 16, 20–23, 29–31, 50, 56, 65–66, 95, 104, 144, 147, 183, 192, 214, 227–228, 233, 236, 240, 242, 245–246, 248–249, 252
empirical sovereignty, 32, 46, 51, 64, 233
encadrement, 17–18, 29, 65, 68, 160, 192–193, 228, 237, 248
enemy-centric COIN, 7, 9–10, 12–13, 43, 46, 62, 65, 80, 137, 139, 171, 175, 214–215, 230, 239
ethnic minority, xvi, 57–58

Europe, 27, 39–40, 43, 52, 55–56, 77, 208
evolutionary change, 117–118, 120, 160, 240

Fall, Bernard B., 18, 33, 35, 190, 223
Fallujah Brigade, 82
Fallujah, Iraq, 76, 81–83, 105, 118–119, 126
Farrell, Theo, 25, 36
faux democracy. *See under* democracy
fedayeen, 80
Female Engagement Teams, xv, 87
Fidros Square, Baghdad, Iraq, 75
Field Manual 27-5 (FM 27-5), *Military Government and Civil Affairs (1940)*, 247
Field Manual 3-24 (FM 3-24), *Counterinsurgency* (2006), 2, 33–35, 71, 86–88, 127, 253
field-grade officers, 155–158, 202
56th Army of the Republic of Vietnam (ARVN) Regiment, 213
Filipino, xv–xvi, 22, 39, 42, 47–48, 50, 53–55, 59–60, 67, 69, 71, 73, 95, 234–235
    Filipino communist party, 61
    Filipino counterinsurgents, 90
    Filipino Nationalists, 41, 43–46, 49, 51–52, 56–57, 61, 64–66, 68
Finlayson, Andrew R., 175, 189, 219–220, 223
1st Battalion, 4th Marines, 83
1st Marine Expeditionary Force (I MEF), 82
force protection, 79, 243
Force X, Philippine military, 63
Foreign Assistance Act of 1974, 15
Foreign Service Institute, 187
foxhole strength (Krepinevich and Linn), 48, 89, 142, 176–177
Franks, Tommy, 77, 79, 245
France, 11, 32, 35, 52, 55, 59, 74, 170, 172, 175, 184, 193, 202, 237, 248–249
French colonial military culture, 216
French colonial model, 185, 194
French imperialists, 39
Frente Sandinista, 162–163
Funston, Fredrick, 45

Galula, David, 2, 8, 32
Garner, Jay, 78–79
gated community, 86
gendarmes or gendarmerie, 59
General Field Order # 8, 53–54
General Order 100, 12, 46, 139
General Order 20, 54
General Order 43, 54
German or Germany, 52, 143, 180, 183, 241
Ghurkas, 55, 73
Giap, Vo Nguyen, 173
Global War on Terror, 244
good governance, 1, 8, 137, 237, 240
Goode, Steven, 13, 34, 71
Goodpaster, Andrew, 202
governance, 1, 7–8, 16–17, 19, 21–23, 27, 29–31, 34–35, 43, 46, 50–53, 60, 64–67, 72, 88, 95–96, 101–104, 120–121, 126, 137, 144, 168, 183, 187, 189–191, 214–215, 227–228, 233–234, 236–237, 240, 243, 245–246, 248, 253–255
graft, 137, 146, 148, 150, 162, 236
Gravatt, Brent L., 4, 32, 35, 147, 155, 161, 163–166
Grayson, William, 43

Green Zone, 96–97, 100–103, 121
Group 559 (North Vietnamese), 169
Guam, 144
Guardia Nacional De Nicaragua, 32, 35, 133–134, 137–139, 141–146, 148–166, 236–237
Guatemala, 180
Guerilla Parties, 12, 46
guerilla warfare, 70, 128, 137–138, 254
Gulf of Tonkin Resolution, 167
Gwynn, Charles W., 2, 8, 32

Haiti, 29, 141
Hama Model, 9
Hamlet Evaluation System (HES), xv, 175, 215, 219–220, 230, 253
hamlets, xv, 14, 94, 174–175, 177, 187, 219–220, 230, 253
Hanoi, North Vietnam, 169
Harrison, Francis Burton, 55
Hay, John, 54
Helmand Province, xvi, 3, 36
Herbst, Jeffery, 249, 255
high embedded strategies, 16, 22, 29, 31, 56, 144, 186, 228, 233, 240–250
Hillah. *See under* Iraq
Ho Chi Minh, 10, 169
Ho Chi Minh Trail, 169
Hoi Chanh (amnestied Viet Cong), 188–189
Honduras, 142–143
Hong Kong, 41, 73
Honolulu Conference, 192
host nation
  host-nation adaptation, 25
  host-nation governance, 16, 22–23, 30, 51, 144, 168, 233, 240
  host-nation government, 1, 7–8,

host nation
  host-nation government (*continued*), 11, 18–20, 23–24, 29, 65, 68, 228, 230, 233, 236–239
  host-nation institutions, 8, 10, 16–17, 21, 26, 238, 240–241, 246, 251
  host-nation security force, 18–20, 23, 29, 65, 68, 162, 214, 228, 230, 236–237, 240–241, 249
  host-nation sovereignty, 4, 16, 18, 29, 234, 236, 247
Hue City, South Vietnam. *See under* Vietnam
Hukbalahap bong Mapagalayang Bayan, Hukbalahaps, 46
Hukbalahap Rebellion, 50, 61–66, 68
humanitarian assistance, 78, 154, 189
humanitarian imperialism, 244, 254
Hussein, Saddam, 75, 78, 80
hybrid COIN methodologies, hybrid-COIN strategy, 44, 239

illustrados, xv, 42, 44, 52, 61, 66
Imam Ali Mosque, 83
imperialists, imperialism, 39, 43, 45, 144, 183, 243, 245, 247–248, 250, 254–255
imperialistic behavior, 244
Improvement and Modernization (I&M) plan, 197
incumbent (executive, party, regime), 135
indentured serfs, 150
independence, 40–42, 51–52, 55–56, 64, 68, 134, 194, 233–234, 249
Indochina, 35, 201, 248–249

Indochina (*continued*)
   Indochinese colonial forces, 55
infiltration rates, 179
influence, 18, 62, 120, 123, 137, 140, 143–144, 158, 160, 183–184, 194, 215–216, 231, 236, 238, 240, 251
   influencing strategy, 19, 237
Institute for Military Support to Governance (IMSG), 246
institution-influencing strategies, 16, 18, 24, 29, 66, 95, 104, 120, 183, 214, 232, 237, 245
institution-inhabiting strategies, 16–17, 26, 29, 51, 60, 65, 144, 193, 237, 242–243, 245, 247–249
insurgents, xv, 1, 12, 24, 26, 28, 30, 46–48, 50, 54–55, 58–63, 71, 81, 83, 89, 101, 108, 113, 115, 119, 122, 139–140, 146, 153–154, 157, 159, 176, 185, 188–189, 206, 233, 238, 241
   insurgency, xvii, 6, 8–11, 15, 17, 25, 27, 32, 40, 45, 52, 64–65, 74, 78–80, 82, 86–88, 102, 141, 158, 170–171, 174–175, 180, 214
   insurgent propaganda, 82
   insurgent sanctuaries, 7, 13–14, 49, 66, 92, 142–143, 178–180, 227, 230–232, 250
   insurgent staging bases, 178
intelligence, xvii, 5, 9, 14, 42, 46, 61, 86–87, 94, 163, 167, 189, 195, 218, 220–222, 224, 232, 253
International Monetary Fund (IMF), 19
Iran, 35, 92
Iraq, 1, 8, 12, 25, 29–30, 33–35, 40, 48, 51, 66, 70, 73, 85, 126–127, 230–233, 239, 246
   Baji, 76, 119

Iraq (*continued*)
   Basra, 96
   Hillah, 100, 125
   Government of Iraq, 77–79, 88–90, 93–96, 101–104, 118, 120–125, 234–235
   Interim Iraqi Government (IIG), 97–99, 121, 234
   Iraq War, xvi, 76, 81, 84, 86, 89, 91–93, 95, 245, 254
   Iraqi, xv–xvii, 2–3, 15, 19, 75–82, 84, 87–90, 92–125, 128–131, 234–236, 242, 245, 251, 253
   Iraqi Army, 3, 77–78, 94, 104, 106, 117–119, 125, 130, 235
   Iraqi Constitution, 84, 95, 97–98, 100, 106, 129, 236, 253
   Iraqi forces, 75–76, 87, 95, 106–109, 112–113, 115, 118–119, 122–123, 235
   Iraqi Governing Council (IGC), 97–98, 234
   Iraqi Kurds, 97, 99–100, 103
   Iraqi police, 77–78, 87, 104, 119
   Iraqi security forces, 76–79, 84, 89, 104–109, 113, 118–120, 122–123, 235, 245
   Iraqi Shia, 99
   Iraqi soldiers, 75–76, 104, 106, 119, 131, 235
   Mosul, 76, 94, 102, 119–120, 125, 235
   Tikrit, 76, 119
Islamic, xv, 9, 61, 65, 79–80, 245
   Islamic insurgency, 235
   Islamic State of Iraq and Syria (ISIS), 2, 15, 34, 76, 89, 92–94, 102, 107, 109, 113, 118–120, 122, 125, 127, 129, 131, 232, 235, 253
isolationism, 250

Japan, 1, 32, 39, 43, 52, 55, 143, 183
Jawad, Saad, 95, 97–98, 100, 129, 253
Jayish Al-Mahdi (JAM), 81, 83
Jefe Director of the Guardia Nacional, 148, 151, 155, 157–158, 165
Johnson, Lyndon B., 192, 194
Joint Security Stations (JSSs), 87
Jones Bill of August 1916, 40
Jones, Frank L., 186, 219
juridical borders, 249
juridical territory, 104, 233

kafil, 96, 121, 234, 253
Karnow, Stanley, 32, 35, 40, 45, 67–68, 70–74
Keane, Jack, 15
Keimling, 152, 165
Keimling, Herbert S., 152, 165
Khanh, Nguyen, 170, 183
Kien Hoa, South Vietnam. *See under* Vietnam
Killcullen, David, 2, 33
Kipling, Rudyard, 40, 70
Kitson, Frank, 2, 8, 32
kleptocratic, 24, 134, 139, 161–162, 237
Kobbe, William A., 54
Komer, Robert, 171, 173, 187, 219
Kosovo, 246
Krepinevich, Andrew, 2, 24, 32, 36, 48, 71, 89, 127, 142, 164, 170, 176, 179, 202, 208, 218–225, 231, 253
Kulak Revolution, 10
Kuwait, xvi, 79

Lansdale, Edward, 63
Laos, 74, 169, 178, 180, 185, 200, 203, 232, 237

Latin America, 35, 140, 158, 163
latitudinal comparison, 28, 227
Lawton, Henry Ware (Major General), 53–54, 57
Lee, Robert E., 159
legal mandate, 145, 244
legislature, 51, 55, 61–62, 99–100, 234
Lend-Lease Agreement, 143
Lewis, Michael, 34, 71, 122, 131, 217
Liberal Generals (in Nicaragua). *See under* Nicaragua, 136
liberal or liberal party (in Nicaragua). *See under* Nicaragua
Liberia, 246–247
Lieber, Francis, 46, 71
Lincoln, Abraham, 12, 159
lines of communication (LOCs), 79, 81, 83, 87
lines of effort, 88, 127
lines of operation, 33, 88, 127
linguistic imperialism, 244, 254
Linn, Brian, 48, 50, 70–73, 89, 127, 142, 164, 176
Lioness Program, xv–xvi, 87
little brown brothers, 67
local governing councils, 3, 44, 51, 88, 96, 121, 129, 234
logistics, xvii, 14, 46, 86, 94, 105–106, 108, 117–118, 123, 127, 148, 153, 168, 177, 181–182, 189–195, 200–201, 217, 232, 238
logistical network, 169, 213
logistical support, 179, 214, 237
London School of Economics, 95
low embedded strategies or strategies of low embeddedness, 16, 23, 183, 214, 228, 233, 236, 240–241, 243–244, 249–250, 252
Luciani, Giacomo, 36, 96, 129, 253
Luna, Antonio, 41

Luzon. *See under* Philippines

M1A1 Abrams main battle tank, 75
Mabini, Apolinario, 42, 53, 66
Macabebe Scouts, xvi, 45, 49, 56–58, 60, 68, 235
MacArthur Jr, Arthur, 44, 46, 58
MacArthur, Douglas, 1, 61, 68
Mack, Andrew, 10, 33
magistrates, 44, 51, 53, 234
Magsaysay, Ramon, 63–64, 68
Mahdavy, Hossein, 23, 35–36, 253
Mahdi Army, 81
Malaysia, 73
  Malaysian Emergency, 59
Mali, 4
Managua, 163, 165
  American Legation in Managua, 147
  Managua Metropolitan Police Force, 147
  Managua Police, 152
  Managua police chief, 152
Manila. *See under* Philippines
Mansfield, Mike, 168, 218
Mao Tse-tung, 7, 10, 44, 70, 128, 169, 242, 254
Mao's Phases of guerilla forces, 92, 169
Marine Corps Legation (embassy) Guard, 134–135, 147–148
Marines, xv–xvi, 2–3, 12, 32, 35, 75, 82–83, 105, 125–126, 133–135, 137–139, 141–143, 145–158, 160–166, 169, 188, 205, 219, 236
Marjah, Afghanistan, xvi, 2–3
martial law, 44, 139, 141, 164
Marxist critiques, 183
mass resettlement, 45
Mattis, James, 2, 12, 15, 32, 251
McCoy, Frank, 146

McGrath, 13, 34, 48–49, 71, 89–90, 127, 176, 231, 253
McGrath, John, 13, 34, 48–49, 71, 89–90, 127, 176, 231, 253
McKinley, William, 41–43, 51–52, 67
McNamara, Robert, 192
Mecklin, John, 184, 222
Mekong River Delta. *See under* Vietnam
Merom, Gil, 9, 11, 33
Merritt, Wesley, 41
Mexico, 136, 243
microsectarian, 235
Migdal, Joel S., 27, 36
military
  military adaptation, 16, 36
  military aid, 63, 167, 182, 184
  Military Assistance Command Vietnam (MACV), 169, 171, 173
  military culture, 26, 123, 216, 235
  military governance, 17, 65, 243, 248
  military government, 15, 17, 35, 51, 55, 234, 241, 243, 246–248
  Military Regions in South Vietnam. *See under* Vietnam
Miller, Paul D., 22, 35
Miller, Stuart Creighton, 47, 67, 71, 74
Millet, Richard, 165
minimum requirement for increasing SLAW, 228, 240
Ministerial Assistance Teams, 19, 101
ministerial assistance teams (MATs), 19, 101
Moncada, José María, 154–156
Montague Jr., Robert M., 183–184,

Montague Jr., Robert M.
(*continued*), 190–191, 222–224, 254
moral hazard, 24, 124, 217, 240
Moro Muslims, 58
Mosul. *See under* Iraq
Moyar, Mark, 175, 220
Multi-National Security Transition Command-Iraq (MNSTC-I), 105
Munro, Dana Carleton, 137, 145, 147, 165–166
Munro-Cuadra Pasos Agreement, 145
Muslim, 79–80, 245
　Muslim Brotherhood, 9
　Muslim insurgencies, 61, 65
mutual adaptation, 25
mutual competitive adaptation, 26

Nagl, John, 2, 25, 32, 36, 88, 127, 219–224, 253–254
national assembly, 42, 100
National Liberation Front (NLF), 169
National Security Council (NSC), 77, 177, 193, 218, 221, 223, 253, 255
National Security Council Assessment 1972, 177, 193, 198, 201, 220–221, 223, 253
National Security Strategy (2012), 250, 255
NATO Training Mission-Iraq (NTM-I), 105
Neimeyer, Charles, 142, 147, 163–166
neo-colonial, 183, 243–244, 247
neo-imperial, 243–245, 254
neo-trusteeships, 15, 34, 243–244, 249, 251, 255
Nicaragua, 23, 29–30, 32, 133,

Nicaragua (*continued*), 138–139, 141, 143, 151–156, 161, 163–166, 230–231, 237–239, 242, 251
　Army for the Defense of Nicaraguan Sovereignty (Ejército Defensor de la Soberanía de Nicaragua-EDSN), 140
　conservatives or conservative party (in Nicaragua), 134–137, 140, 146, 157–158, 162
　Liberal Generals, 136
　liberal or liberal party (in Nicaragua), 134, 136–137, 140, 145, 157–158, 162
　liberal uprising in 1926, 136
　Nicaraguan Army, 145, 149–150
　Nicaraguan government, 135–136, 144, 146–148, 159–160, 236
　Ocotal, 145
　San Juan Del Norte, 134
Niger, 4
Ninh, Tay, 175, 220
Nixon, Richard M., 194
nongovernmental organizations (NGOs), 244
North Korea, 61, 193
North Vietnam. *See under* Vietnam
Northrop Grumman, 104
nuclear war, 169
Nueva Segovia, Nicaragua, 140

Obama, Barack, 76
Observe, Orient, Decide, Act Loop (OODA Loop), 25
occupation, 34, 43, 71, 79, 96, 98, 125, 141, 227, 241, 243, 245, 247–248

Ocotal region. *See under* Nicaragua
Odierno, Raymond, 15
offensive air support, 95, 232, 238
Office for Reconstruction and Humanitarian Assistance (ORHA), 78
oligarchical kleptocracy, 134, 161–162
oligarchy of intelligence, 42
101st Airborne Division, 169
Operation Al Fajr (the Dawn), 81–82
Operation Attleboro, 172
Operation Cedar Falls, 172
Operation Harvest Moon, 172, 219
Operation Hastings, 172
Operation Junction City, 172
Operation Lam Son, 200
Operation Linebacker, 213
Operation Long Reach, 172
Operation Masher, 172
Operation Moshtarak, 3
Operation Restoring Rights, 106
Operation Starlight, 172
Operation Steel Curtain, 83
Operation Van Buren, 172
Operation Vigilant Resolve, 82, 105
Operational Plan (OPLAN) El Paso, 179
operational security (OPSEC), 109
Osmena, Sergio, 40
Otis, Elwell Stephen, 41, 44, 46, 53–54, 57, 72
Ottoman government, 245
outside-in (methods of development and effecting change), 121, 183, 217, 236, 238

Pacific Squadron, 39
pacification, xv, 71, 83, 173–175, 177, 186–188, 190, 198, 215, 217,

pacification (*continued*), 219, 222, 230, 232, 237
Pact of Espino Negro, 136, 158
padrone, 140
Pampanga, the Philippines, 56
Panama, 28, 141, 144
Panetta, Leon, 15
Paris Peace Accords (1973), 214
Paterno, Pedro, 44
People's Liberation Army, 61
People's Republic of China, 182
Pershing, John J., 59
Petraeus, David, xvii, 2, 15, 34, 84, 86–88, 105–106, 127–128, 253–254
Philippines, xv, 1, 5, 12, 23, 28–30, 32, 40, 46, 48, 50, 52, 57, 65, 67–71, 74, 89–90, 95, 101, 127, 141, 144, 163–164, 231, 233, 239, 242–244, 251
 Luzon, 45, 49, 56, 61, 66
 Manila, 39, 41–43, 51, 53–55, 60
 Philippine Constabulary, 59–60, 73
 Philippine Dictatorial Government, 41
 Philippine judiciary, 44, 54, 234
 Philippine legislature, 51, 55, 61–62
 Philippine military, 51, 56, 59–60, 63, 235
 Philippine Military Academy, 56, 60
 Philippine Scouts, xvi, 49, 56, 58–60, 72–73
 Philippine security forces, 56, 60–62, 64, 235
 Philippine Supreme Court, 51, 54
Philippine-American War, 35, 43, 47, 139

Index   269

Phoenix Program (Phung Hoang), 174–175, 189, 205, 220
Phouc Long, South Vietnam, 171
policy of moderation and reprisal, 12, 63
political ideology, 140
political tenability, 242–243
political will, 24, 242
politics by other means, 242
popular uprising, 179
population-centric COIN, 2, 7–10, 12–13, 43, 46, 65, 83–84, 92, 139, 141, 171, 174–175, 215, 230, 239
Powell-Weinberger Doctrine, xvi, 8
President Kennedy, 185, 194, 237, 246
Presidential Envoy to Iraq, 78
pretense of sovereignty, 144, 185–186, 234
probability of success, 242, 245
professional military education (PME) system, 248
protraction, 138
provincial councils, 54, 88
provincial level, 102, 174, 191, 237
Provincial Reconnaissance Unit (PRU), 175, 205
Provincial Reconstruction Teams (PRTs), 102–103, 129
Puerto Rico, 144, 243
Puller, Lewis Burwell "Chesty", 141

Quang Ngai, South Vietnam. See under Vietnam
Quang Tri province, South Vietnam. See under Vietnam
Quantico, 2, 34, 219
Quezon, Manuel, 40, 47, 55
Quinlivan, James, 13, 34, 71, 89, 176, 231, 253
Quirino, Elpidio, 62

racial prejudice, 140
Ramadi, Iraq, 2–3, 76, 83, 88, 94, 96, 118–119, 125, 129
Rand Corporation, 77, 79, 125
reconstruction, 76, 78–79, 96, 101–104, 129, 243
Reconstruction Management Office, 101–102
Record, Jeffrey, 11, 33
Regimental Combat Team 1 (RCT 1) (USMC), 82
Regimental Combat Team 7 (RCT 7) (USMC), 75, 82
Regional Forces/Popular Forces (RF/PFs), 187–189, 202, 206
rentier state, 23–24, 26, 35–36, 75, 96–97, 120, 124, 129, 212, 214, 217, 228, 232, 240, 250–251, 253
Republic of Vietnam Armed Forces (RVNAF), 168–169, 174, 176–178, 180–182, 185, 187, 192–203, 205–208, 212–217, 223–224, 238
resettlement program, 141
retarding effect, 122, 217, 241
revolt, 33, 55–56, 74, 134, 136–137, 139–140, 148, 161
revolutionary timeframe, 120, 160, 228, 240
Rhineland 1918–1923, 243
Roosevelt, Theodore, 45, 147, 166
Root, Elihu, 42–43, 51
Rough Riders, 147
Roxas, Manuel, 62
Royal Malaysian Special Branch, 59
Rubin, Barnett R., 24, 36
Russia, 92

Sacasa, Juan Bautista, 135–136, 157–160

Saigon, South Vietnam. *See under* Vietnam
San Juan Del Norte. *See under* Nicaragua
sanctuary, 7, 13–14, 49, 65–66, 92, 142–143, 178–180, 227, 230–232, 250
Sandinistas, 133–134, 137–139, 141–143, 145, 155, 158–164, 180
Sandino, Augusto Cesar, 136, 138–143, 146, 149, 157–161, 163
Saudi Arabian army, 104
School of Military Governance, 248
Schurman Commission, 53
Schwarz, Rolf, 24, 36
Scull, Greg, 147
search-and-destroy, 171–172
2nd Marine Brigade in Nicaragua, 138, 142, 146, 151
2nd Philippine Commission, 59, 72
Sectarian insurgent groups, 80
sectarian strife, 56, 58, 60, 68, 80, 84, 89, 235
security, 7–8, 21–22, 25, 31–33, 35, 50, 66, 81–82, 85, 87, 94, 96–100, 102, 121, 125–130, 133, 144, 148, 152, 154, 159, 161, 174, 176–177, 183, 186–189, 191–193, 218, 221, 223, 230, 238–239, 246, 249–250, 253–255
    security cooperation, 19, 248, 251
    security forces, 1, 16, 18–20, 23–24, 29, 56, 60–62, 64–65, 67–69, 76–79, 84, 88–90, 93, 95, 104–109, 113, 118–120, 122–124, 136–138, 146–147, 150, 160, 162, 214, 228, 234–237, 240–241, 245, 279
    security institutions, 60–61, 160, 240, 245

shared sovereignty, 15, 17–18, 34–35, 51, 53, 72, 234, 243, 246, 255
Sheik Satar Albu Risha, 88
Shia, 80–81, 83–84, 88, 92, 99–101, 103, 118, 121–122, 125, 235
Sierra Leone, 18, 246–247
significant acts (SIGACTS), 3, 84–85, 89
Sinjar, Iraq, 94
Small Wars Manual, The, 34–35, 247
Smith, Julian C., 141, 163–165
socialist critiques of colonialism and imperialism, 183
Solórzano, Carlos, 135–136
Somoza, Anastasio, 143, 158–161
South Vietnam. *See under* Vietnam
sovereignty, 4, 15–18, 29, 32, 34–35, 43, 46, 51–53, 55–56, 58, 64, 72, 99, 101, 121, 140, 145, 186, 233–234, 243, 246–247, 255
sovereignty fallacy, 144, 146, 183–185, 236
Soviet Union, 27
Soviet trainers/advisors, 94, 182
Soviet War in Afghanistan, 80
Spain, 41–42
    Spanish American War, 60
    Spanish Empire, 39
    Spanish Mestizos, 140
    Spanish Pacific Squadron, 39
Special National Intelligence Estimate 53-64 (SNIE 53-64), 167–168
spoiling attacks, 180, 231–232
stability, 35, 64–65, 77, 79, 85, 126–130, 135, 147–148, 160–161, 163, 236, 279
    stability operations, 34, 71, 251, 253
staging bases, 178–180, 232

Index

Stalin, Joseph, 10
starvation, 45, 136
state failure, 16–17, 35, 143, 239, 249–250, 255
state longevity after withdrawal (SLAW), 6–7, 9, 11, 13–17, 20–22, 27–32, 47, 49, 65, 89, 144, 180, 215, 227–233, 236–237, 240, 242, 246, 248, 250, 253
state-centric COIN theory, 16, 27–28, 233, 238, 240
  implications of state-centric COIN theory, 227
state-to-state warfare, 180
Status of forces agreement (SOFA), 104
Stimpson Peace Plan, 136
Stimson, Henry L., 136, 145, 151, 156–158
strategies
  less invasive strategies, 244–245
  strategy of attraction and chastisement, 65
  strategic rentier state, 23–24, 26, 75, 120, 124, 212, 214, 217, 228, 232, 240, 250–251
  strategic rents, 24, 120, 124, 181, 212, 214, 217, 228, 232, 240, 251
  strategies of high embeddedness. *See* high embedded strategies
  strategies of low embeddedness. *See* low embedded strategies
strongmen, 241
stumbling state, 22–23, 39, 65, 68–69, 251
Sultanate of Sulu, 49, 58–59
Sulu Archipelago, 61
Summer Stalemate, 50
Summers, Harry, 14, 34, 178–179,

Summers, Harry (*continued*), 221
Sunni Arabs, xvii, 80–82, 84, 88–89, 99–101, 103, 118, 122, 125, 235
Sunni tribal militias, 118
support after withdrawal, 50, 93, 143, 180, 227–228, 232
surge, xvii, 2–3, 80, 84, 86, 89–90, 92, 103–104, 107, 126–128, 189, 230
symbiotic adaptation, 17, 26, 60, 68, 153, 240
Syria, 4, 9–10, 76, 92, 128, 235
  Damascus, 245

tabula rasa governance, 29–30, 105, 227, 233–234
Taft Commission, 44–45, 53
Taft, Willian Howard, 44–45, 53, 55
Tagalog, xvi, 57–58, 235
Taji Prison, Iraq, 118
Tal Afar, Iraq, 83, 106, 130
Talabani, Jalal, 100
Taliban, 2–3, 80
Tammany Hall system, 67
targeted raids, 87
Taruc, Luis, 61, 64
Taylor, Maxwell B., 195
Teller Amendment, 43
Temporary Equipment Recovery Mission (TERM), 194
territorially associated units, 185
Tet Offensive, 171, 173–175, 177, 197, 215
theater-security cooperation, 19, 248, 251
third-party COIN, 10–11, 16, 22, 26–28, 30, 230, 238–240
third-party counterinsurgents, 238
Thockmorton, John L., 192–193
Thompson, Sir Robert, 8, 32, 188

Thu Duc Officer Candidate School, 202
Tikrit. *See under* Iraq
top-down transitions (governance and security), 95, 121, 234–235, 241
training effect, 122, 217, 241
transisthmian wireless communication, 153
Transitional Administrative Law, 97–99
Treaty of Paris, 41–43, 51
Trinquier, Roger, 2, 8, 32
troop-to-population or counterinsurgent-to-populace ratio, 7, 13, 30, 47–49, 65, 86, 89–92, 141–143, 176–178, 227, 230–231, 239, 250
troop-to-terrain ratio, 142
Truman, Harry, 1, 62
trusteeship, 15, 17–18, 34–35, 40, 51–53, 55, 66, 68, 234, 243–244, 246, 248–249, 251, 255
tumbling state, 22–23, 133, 160, 251
Twentynine Palms, California, 2

Ucko, David, 9–10, 33
UN Security Council Resolution (UNSCR) 1483, 79, 97, 125
unilateral adaptations, 240
universal suffrage, 42
US Agency for International Development (USAID), 77, 128
US Air Force, 168, 173
US Army, 2, 35–36, 56, 77–78, 81, 107, 126–127, 145–146, 148, 163, 172, 174, 178–179, 186, 188, 197, 203, 212, 222–223, 225, 236, 243, 246–248
US Army Command and General Staff College, 36, 212, 223, 225

US Army John F. Kennedy Special Warfare Center and School, 246
US Army mobile training teams (MTTs), 188
US Army War College, 186, 222
US Civil War, 46, 63, 86, 134, 139, 159
US Congress, 43, 55, 167, 237
US Department of State's Civil Response Corps (CRC), 248
US Embassy in Saigon. *See under* Vietnam
US Information Services, 184
US Joint Chiefs of Staff, 62
US Marine Base at Khe Sanh, 173
US Marine Corps, xv, 2–3, 12, 14, 34–35, 70, 75, 82–83, 105, 107, 126–127, 130, 133–134, 137–139, 141, 143, 145, 147, 149–152, 160, 163, 169, 189, 194–195, 198, 200, 203, 205–206, 219, 236, 243, 248
US Marine Corps University, 122, 217
US Marine Corps' Combined Action Platoons (CAP), 173–174, 186, 188
US Military Academy, 56, 60
US military advisors, 237
US Navy, 145–146, 153
US Navy Special Service Squadron, 146
USS Cyane, 134

Van Creveld, Martin, 9
Vann, John Paul, 187
VC/NVA 5th Division, 213
Veracruz, Mexico, 243
Viet Cong, xvii, 14, 170–171, 173–174, 176–177, 179, 187, 195, 197–198, 205–207, 213, 215, 217–218, 220–221, 230–232

Viet Cong (*continued*)
  Viet Cong Infrastructure (VCI), 175, 188–189
  Viet Cong main force threat, 172
  Viet Cong main force units, 169
  Viet Cong propaganda, 183
Viet Minh, 172
Vietnam, 2, 8, 11, 30, 32, 36, 71, 74, 92, 124, 127, 143, 164, 188, 239, 245–246, 254
  Army of the Republic of Vietnam (ARVN), 168–173, 181, 185, 189, 193–194, 198, 200, 202, 206, 212–213, 217
  Danang, 169
  Government of (South) Vietnam (GVN), 14–15, 168–171, 173, 175–178, 180–186, 189–191, 197, 200–203, 208, 212, 214–217, 232, 237
  Hue City, 173
  Kien Hoa, 169
  Mekong River Delta, 170
  Military Regions in South Vietnam, 213
  North Vietnam, xvii, 14–15, 168–169, 171, 175, 178–182, 185, 189, 193, 197–198, 200–203, 205–206, 213–216, 218–221, 223–225, 232, 253
  North Vietnamese Army (NVA), xvii, 14, 171–174, 176–177, 180, 185, 187, 194–195, 197–198, 202, 206–207, 212–214, 217–221, 225, 231–232
  North Vietnamese Army (NVA) invasions (1970, 1971, 1972 & 1974–1975), 180, 194, 212–213
  Quang Ngai, 171

Vietnam (*continued*)
  Quang Tri province, 213
  Saigon, 14, 173, 175, 181, 184–185, 189, 214, 237
  South Vietnam, xv, xvii, 3, 5, 14–15, 24–25, 29, 72, 167–171, 173–187, 189–203, 205–206, 208–225, 230–232, 236–238, 242, 253
  South Vietnamese command and general staff course, 202
  US Embassy in Saigon, 173
  Vietnam Training Center (VTC), 187
  Vietnam War, 34, 86, 218, 220, 224
  Vietnamese military academy, 202
  Vietnamese officer commissioning programs, 202
  Vietnamese Rangers, 205
  Vietnamization, xvii, 197
Vinell, 104–105

Wade, Christine, 139–140, 162, 164, 166
Walker, Thomas, 139–140, 162, 164, 166
Washington DC, 33–35, 37, 70, 72–73, 125, 127–128, 130–131, 164–165, 185, 196, 199, 218–221, 254–255
wealth inequalities, 140
Westmoreland, William, 169–175, 179, 187–188, 192–193
whole-of-government approach, 187, 189
Wiwili, Nicaragua, 160
Wolff, Leon, 47, 71, 74
World War II, 18, 35, 46, 50, 61,

World War II (*continued*), 73–74, 143–144, 180, 183, 242–243, 248
World War I, 17, 59, 138

Zelaya, Jose Maria, 158
Zinni, Anthony, 1, 14–15, 32, 34, 168, 178, 205–206, 218, 221, 246–247, 254–255

Cambria Rapid Communications in Conflict and Security (RCCS) Series

General Editor: Geoffrey R. H. Burn

The aim of the RCCS series is to provide policy makers, practitioners, analysts, and academics with in-depth analysis of fast-moving topics that require urgent yet informed debate.

Since its launch in October 2015, the RCCS series has the following book publications:

- *A New Strategy for Complex Warfare: Combined Effects in East Asia* by Thomas A. Drohan
- *US National Security: New Threats, Old Realities* by Paul Viotti
- *Security Forces in African States: Cases and Assessment* edited by Paul Shemella and Nicholas Tomb
- *Trust and Distrust in Sino-American Relations: Challenge and Opportunity* by Steve Chan
- *The Gathering Pacific Storm: Emerging US-China Strategic Competition in Defense Technological and Industrial Development* edited by Tai Ming Cheung and Thomas G. Mahnken
- *Military Strategy for the 21st Century: People, Connectivity, and Competitipauon* by Charles Cleveland, Benjamin Jensen, Susan Bryant, and Arnel David
- *Ensuring National Government Stability After US Counterinsurgency Operations: The Critical Measure of Success* by Dallas E. Shaw Jr.

For more information, visit www.cambriapress.com.

www.ingramcontent.com/pod-product-compliance
Lightning Source LLC
Chambersburg PA
CBHW032033300426
44117CB00009B/1038